Andrew Dart

Building Your Skeptical Toolkit
A Beginners Guide to Science, Skepticism and Critical Thinking Skills

Building Your Skeptical Toolkit
Copyright © 2012 Andrew Dart

First Published: 2012
Publisher: RabbitPirate Publishing

The right of Andrew Dart to be identified as author of this Work has been asserted by him in accordance with sections 77 and 78 of the Copyright, Designs and Patents Act 1988.

All rights reserved. No part of this publication may be reproduced, distributed, or transmitted in any form or by any means, including photocopying, recording, or other electronic or mechanical methods, without the prior written permission of the publisher, except in the case of brief quotations embodied in critical reviews and certain other noncommercial uses permitted by copyright law.

ISBN: 1479328383
ISBN-13: 978-1479328383

Visit **www.andrewdart.co.uk** for more information on science, skepticism and critical think and to see other books by this author.

For Andy and Abby

For putting up with me talking about this stuff late into the night on more than one occasion. Your suffering was appreciated.

Acknowledgements

This book never would have been finished if not for the constant support and assistance of my friends and family. My thanks go out to my good friends Andy, Abby, Shane, Dave and Lauren for acting as proof readers and to my Mum for acting as my editor and being my biggest supporter. Any mistakes still left in this book are my doing not theirs. I also want to thank AndromedasWake, Th1sWasATriumph and AngryWomble for helping to keep the fires of skepticism burning in me and for keeping me going at a time when I was thinking of scrapping the whole thing, thanks guys. Special thanks go out to Ellen for all of the many times I have kept her waiting whilst I have been working on this book. Sorry about that. And lastly I want to thank you for reading this book, you have no idea how much that means to me. Thank you.

Contents

Introduction – Why is Skepticism Important? 1

PART ONE – Science! It works, bitches!

So what's this all about then? – What skepticism is and what it is not 9

Skeptical Sidebar #1 – The behind the back mentalism trick 17

Just the facts Ma'am – Facts, Hypothesis, Theories and Laws 27

Method in my madness – A quick look at the scientific method 39

A law unto themselves – The three laws of logic 47

Skeptical Sidebar #2 – Syllogistic reasoning 51

That's highly illogical Captain – An incomplete guide to logical fallacies 55

The Good, The Bad and The Extraordinary – What makes something good evidence? 77

Skeptical Sidebar #3 – The file drawer effect 93

Skirting the razors edge – Occam's razor and Hume's maxim 97

Is there a Doctor in the house? – A quick guide to clinical trials 101

Do you take the red pill or the blue pill? – The placebo effect 109

Stop global warming, become a pirate – Correlation and Causation, Possibility and Probability	115
You cannae change the laws of physics Jim – Energy, entropy and quantum mechanics	125
Time for a little Mad Science – How to identify pseudoscience	139
Open mind or an empty head? – Do skeptics have an open mind?	167
That doesn't make sense – Science and common sense	173
You can't prove I'm wrong – The burden of proof and asking the right questions	179
Going beyond the realm of science – Science and the supernatural	185

PART TWO – Your Brain Hates You!

Tell me something I don't already know – The power of confirmation bias	201
I know it's wrong, but… - Cognitive dissonance	215
I remember when – Past lives, false memories and Satan!	227
Skeptical Sidebar #4 – The amazing memory trick	243
I am not a number… - The power of conformity and the wisdom of crowds	249
Skeptical Sidebar #5 – The gambler's fallacy and randomness	257
What are the chances of that? – The law of large numbers and how we suck at maths	261

Skeptical Sidebar #6 – The Monty Hall problem	275
Let me tell you your future – The less than magical art of cold reading	281
Skeptical Sidebar #7 – The book test	311
Can I have a vowel please Carol? – Devils, dowsing and the ideomotor effect	317
Here's looking at you kid – The wonderful world of pareidolia and patterns that aren't really there	341
I think I'm hearing things – Backmasking, reverse speech and EVP	363
Skeptical Sidebar #8 – Is seeing really believing?	389
Aliens, Witches and Demons, oh my! – A nightmarish look at sleep paralysis	393
Skeptical Sidebar #9 – How far? How high? How big? How fast?	415
And in conclusion... - Beginning your skeptical journey	423
Bibliography and further reading	431

Introduction

Why is skepticism important?

This question is perhaps the most important issue for me to address in this introduction and a great place to start any book on the subject. After all if skepticism is of no import and carries no weight then any information on how to be a better skeptic is itself not worth the paper it is written on, let alone your time and effort spent reading it.

Unfortunately, the answer to this question is not exactly a simple one. It varies from person to person and depends heavily upon your answer to another important question. In his pivotal book on the subject, poetically entitled *The Demon Haunted World*, legendary astronomer and science populariser Carl Sagan spent a large part of his first chapter addressing this underlying question. He broke it down into two similar but distinct parts.

"Do we care what is true and does it matter?"

Positive answers to this two part question form the central tenets of good skepticism. A skeptic, when all else is stripped away, is, at their core, someone who cares passionately about the truth. All the other things that skeptics tend to have in common, such as a love of science, a questioning nature and a fascination with the limits of human knowledge, count for next to nothing if they are not built upon an unshakeable foundation of love for the truth.
The truth can be a wonderful, enlightening and world changing thing. Equally it can be ugly, scary and devastating and it is in these cases that a good skeptic will stand out from the crowd. It is

Introduction

unlikely that anyone, when faced with the question above, would announce that they simply don't care about the truth of a claim, or that the value of things that are untrue is greater than the value of those that are. The issue arises when you take a deeper look at their reasons for seeking the truth in the first place.

Do you seek the truth because you believe it will bring you comfort? Is a feeling of superiority, inclusion or belonging the reason you try to search out the truth? Maybe you enjoy the knowledge that comes with the truth or the respect you get from others who may lack that knowledge? If any of these are the case then you should ask yourself if it is the truth that really matters to you or the feelings you believe will come with that truth?

Great comfort, gratification and camaraderie can be found in all sorts of claims, from the belief that positive thinking will make the World a better place or simply help you lose a few pounds all the way through to the calm that comes with the conviction that your loved ones are waiting for you beyond the veil of death. If a sense of well-being and fulfilment is what you are seeking from life then many a paranormal, pseudoscientific or religious claim can provide just what you are looking for.

However if it is the actual truth of something that is important to you then a good skeptical outlook on life is the way to go. Because at the end of the day there are not always easy answers to life's hard questions. It is entirely possible that there is no overreaching point to our existence and that all that awaits us beyond the grave is eternal nothingness, that and a slow decay back to our component parts. Skepticism cannot promise you happiness, success or even a reason to get up in the morning. For those things you will need to look elsewhere. What it can do, more than any other method yet devised by man, is help you find the truth.

Unsurprisingly then skepticism itself is not an easy answer. It takes work, it goes against our nature and at times it can be a hard and lonely road to walk. Many people see a declaration that you are skeptical of some claim or another as a negative thing or even as some kind of character flaw. They may equate the word skeptic with the word cynic, in fact even the thesaurus in Microsoft Word does this, when in reality nothing could be further from the truth.

Just because you demand a certain level of evidence before accepting a claim doesn't mean that you would not love for it to be true. I am now, much to my chagrin, well into adulthood and I consider myself something of an ardent skeptic. However I know few people my age who wish that magic was real more than I do. There is a part of me that even now hopes to find a portal to another world every time I open my wardrobe. If someone were to come up with evidence, real hard scientific evidence, that fairies exist I would be ecstatic. If they did the same for dragons my head would probably explode from excitement. Just because right now I do not believe that any of these things are at all likely does not for a second mean that I cannot understand the fascination and enjoyment that comes from imagining that they might be.

For me my wonder at the World around us is only enhanced by my skepticism. It in no way detracts from the majesty of life to see natural explanations to the awe inspiring marvel that is our universe where others see the hand of some supernatural agent. One of my greatest joys in life is to simply take a long walk around the small village in which I live and just take a good look at the World. Where some may seek pixies hiding amongst the bluebells I have no problem seeing wonderful and amazing things right before me in the dazzling dance of the bumblebee and the complex and beautiful interaction between plant and insect.

When out I have occasionally come across delicate masses of white thread, like infinitely fine cotton wool, far from any obvious

Introduction

source and which vanishes before my eyes if touched. Could this really be some unearthly material from the engines of UFOs, as some proponents suggest, or perhaps the hair of angelic beings or saintly visitors? These are certainly intriguing possibilities, and yet they fail to inspire in me the same sense of wonder brought about by the idea that they are simply, a word I use loosely, the remains of the silken parachutes of hundreds of juvenile migratory spiders, used to catch the wind and thus enable them to travel distances they could never manage otherwise. There is something intrinsic to this natural explanation that the paranormal answers lack, even if it does send a shiver down this arachnophobe's spine.

For me the truth has a beauty to it that far surpasses anything that fanciful flights of human imagination can come up with, and this comes from someone who dreams of making a career for himself as a writer of fiction. Of course there is more to a skeptical and reasoned approach to life than just seeing the true beauty and majesty that the World, nay the universe, has to offer. Skepticism can also act as a shield against the ever-increasing complexity and confusion of modern life as well as a sword to wield against those who would prey upon your naivety and ignorance for personal gain.

Psychics target the recently bereaved; claiming to offer them the chance to reconnect with their departed loved ones, for a small fee of course. People hand over small fortunes to these people, hanging on their every word as they gain more and more control over their lives. Nigerian email scammers pepper our mailboxes with offers of vast sums of money if we will simply help them out a little and, while most of us see through their lies, every year numerous people worldwide suffer financial and even physical harm because they take these people at their word.

Conspiracy theorists, political ideologues and religious fundamentalists, often fronted by TV celebrities lacking in any scientific training but instead armed with far more public friendly powers such as their "mommy instinct"*, have convinced tens of thousands of people around the globe of an evidentiary and scientifically vacuous connection between vaccines and autism, forced sterilisation and even AIDS. As a result diseases such as mumps, whooping cough and polio, all recognised killers, the last of which was almost eradicated from the World, are now once again on the rise.

Alternative medical treatments, many of which have no evidence to support their efficacy and which have in fact been shown to be dangerous in some cases, are available everywhere and are even advertised by some GP surgeries as an alternative to more main stream treatments. Pharmacies have entire sections dedicated to homeopathic treatments; a form of alternative medicine that relies upon illogical, magical and demonstrably false concepts that were they in fact accurate would require a rewrite of almost all of physics, chemistry and biology. And yet despite this, and the fact that in 2009 a representative of Boots pharmacy, one of the largest pharmacy chains in the UK, openly admitted that they have no evidence that homeopathic treatments are in any way effective, you will find homeopathic treatments on offer for such conditions as childhood diarrhoea, malaria and even cancer, again all recognised killers. You can even get homeopathic treatments for your pets.

But perhaps these examples seem somewhat indirect; worrying yes but at the same time not issues that you feel can result in immediate injury or loss of life. Well then, consider the direct damage that not thinking skeptically has caused in war torn Iraq.

---

* The 'Mommy Instinct' is what anti-vaccination proponent Jenny McCarthy claims allows her to know that her son's autism was caused by the MMR vaccine in the face of overwhelming scientific evidence against this idea.

Introduction

In the city of Baghdad armed checkpoints are constantly on the lookout for illegal fire arms and explosives that could be used to cause injury or deaths amongst both the military personnel stations there and the civilian population. Around 2009 to help them accomplish this goal the Iraqi government purchased, at the price of $16,500 to $60,000 per unit, what amounted to high tech dowsing rods, more than 1,500 of them, devices that the United States military and technical experts have labelled "useless". Already numerous bombing deaths can be directly linked to checkpoints using these pseudoscientific devices and therefore indirectly to the Iraqi government's failure to think skeptically.[*]

It sounds scary and many of these things are genuine reasons for concern, but never fear. Skepticism and critical thinking can equip you with the tools needed to effectively recognise and combat charlatans, pseudoscience and hoaxes of all shapes and sizes. So why is skepticism important? It seems that after all maybe the answer is simple. It's because the truth is important and skepticism is one of the most effective ways we have of discovering and recognising the truth when we find it. The World is full of confusion and unscrupulous people just waiting to take advantage of us. It is also full of truly wondrous hidden beauty that can open up new and vast horizons to you, beyond even the marvel of human imagination.

At the end of the day skepticism will help you find the truth and, as the saying goes, the truth will set you free.

---

[*] On the plus side in 2010 Jim McCormick, the owner of the company producing these "bomb detectors", was arrested on suspicion of fraud and the UK banned export of these devices. Unfortunately, these actions came 275 deaths and £52m too late.

Part One

Science! It Works, Bitches!

A few things every good skeptic should know about science

So what's this all about then?
What skepticism is and what it is not

Before we start delving into the things that every good skeptic should know it is worth taking a few short pages to look at exactly what it means to be a skeptic. I've talked about why I think skepticism is important but let's take a moment to look a little deeper at what exactly it's all about, and probably the easier way to explain skeptics and skepticism is by looking at what they are not.

Firstly I should make it clear that skepticism is not a belief system, nor is it a form of religion or even a philosophy. Skepticism has nothing to do with telling people *what* to think and everything with teaching them *how* to think. As author and prominent skeptic Dr Michael Shermer put it *"Skepticism is not a position; it's a process."* The dictionary describes skepticism as *"A doubting or questioning attitude or state of mind"* and while this certainly forms a part of skepticism I find this an inadequate and even slightly negative definition.

To my mind skepticism is perhaps best described, as the title of this book implies, as a set of tools, honed by time and field-tested by science and academia into the best instruments we have for discovering the truth. These tools come in all shapes and sizes and like all good tools can be applied in many varied situations. Throughout this book we will be looking at some of these tools and how they can be applied.

Skepticism is often mistakenly equated with cynicism. People think that skeptics are just people who disbelieve things, who like to pull the rug out from under peoples most cherished beliefs, who are contrary just for the sake of being so. Debunkers determined to crush any sense of magic and wonder in the World because they themselves believe in nothing. This, as you shall see, is far from the case. Let's break those arguments down.

As I said in the introduction a skeptic is maybe the furthest thing from a cynic there is. Most of us would love magic to be real and if you could provide evidence for some supernatural claim would be over the moon. But we also see unending wonder and beauty in the natural world and are often driven by a passion for understanding and an almost magnetic attraction to mystery. Skepticism has a hunger for knowledge, a hunger fed by rational enquiring, constructive doubt and the questioning nature of an open, inquisitive mind. It allows us to find joy and magic in the majesty of reality and yet still allows our hopeful hearts to be open to the possibility of something more. To claim that a skeptic is the same thing as a cynic is to completely misunderstand *everything* that skepticism is about.

Skepticism is also not about disbelief. These days you often come across people who either describe themselves or others as skeptics on the basis of something they don't believe in. A great example of this are the "global warming skeptics" who label themselves as such based upon their disbelief in global warming. But as we covered above skepticism is not a position, it is a process for finding the truth, and as such this is an incorrect use of the word. Skepticism is about following the evidence wherever it leads even if you don't like where it is going. It is not an ideological or political position and its only allegiance is to the truth. Simple disbelief in some claim is not skepticism, even if the evidence justifies that disbelief.

Skepticism equally has nothing to do with opposing claims just for the sake of opposing them. Enquiry and not opposition is what drives skepticism. It is about examining claims by the light of reason, logic and scientific investigation; it is not about just opposing them. Of course skeptics do oppose many paranormal and pseudo-scientific claims and yet this is not because of some default automatic mechanism built into skepticism itself. They do so because they have examined the claims and the evidence for them and have found them wanting. The default position of skepticism when faced with a new claim is to reserve judgement until it can be adequately examined, it most definitely does not start from a position of outright opposition.

So how about the claim that skepticism is all about debunking? At first glance this may seem to hold some water as debunking is often the end result a good skepticism. But once again this is not what skepticism is about. One of the consequences of seeking the truth is that you will also discover those things that are untrue. Skepticism does not set out with the goal of debunking any claim, however if said claim is nonsense or just plain untrue then a legitimate search for truth will show this to be the case. So it is true that skeptics often do end up debunking certain claims, but then separating out the chaff is simply a natural outcome of critical thinking and skepticism rather than a goal. And anyway is there really anything wrong with revealing nonsense, hoaxes, scams and misinformation for what they really are?

The final claim I want to look at here is that skeptics don't believe in anything. This surprisingly common belief most likely stems from the previously mentioned idea that skepticism is all about disbelief as well as the completely understandable confusion between the methodological approach of skepticism and the philosophical ponderings on the nature of knowledge that make up philosophical skepticism.

As I said before skepticism is a process and not a position and disbelief in something is as much a position as is belief in something. It is instead a set of tools for evaluating the World around us and not a statement of belief, one way or another. Good, well-applied skepticism can lead to either belief or disbelief in a given claim but, as with the debunking claim, this is not the goal of skepticism.

On top of all that the claim is simply untrue. Skeptics believe all manner of wild and wondrous things. Skepticism teaches you how to think and not what to think. As such there are skeptics who, by following this method, have come to a belief in ghosts or Bigfoot. There are many religious skeptics who have come to their beliefs by following the evidence to what they believe to be the most logical and rational conclusion. Skepticism is about the journey and not the destination and while I may disagree with some of the conclusions of a few of my fellow skeptics that does not mean I can truly fault their skepticism. After all I wouldn't be a very good skeptic if I didn't consider the possibility that I may be the one who is wrong.

Which brings us nicely to some of the misconceptions that skeptics sometimes have about themselves and about those who do not apply the tools of skepticism to their beliefs. First and foremost being a skeptic does not make you more intelligent than non-skeptics. That is something that I want you to keep in mind as you read this book. Using the tools of skepticism has nothing at all to do with how intelligent you are. Many very smart people believe in all manner of supernatural and pseudoscientific claims. Being a brain surgeon doesn't mean you automatically won't believe in ghosts or astral protection any more than being a rocket scientist protects you from falling victim to the preposterous claims of homeopathy.

Yes, there is undoubtedly a correlation between level of education and skepticism however this does not mean one requires the other and it is not at all unusual for the highly educated members of society to display appalling skepticism and

Building your Skeptical Toolkit

believe in all manner of hokum while those with little or no education approach the World with the refined and effective critical thinking skills of a seasoned skeptic.

Additionally the smarter someone is then the more effectively they can argue in favour of their erroneous beliefs. Never assume that because someone believes in alien abductions that they will be a push over in debate or that you are the most intelligent person in the room just because the person across from you chooses to consult a psychic. Many a prideful skeptic has faced their undoing at the hands of a well-educated and intelligent believer in the paranormal.

Also, just because you are a skeptic don't ever assume that you can't be fooled. Skeptics are just as susceptible to being taken in as anyone else. The only difference between you and a non-skeptic is that you have a set of tools for effectively evaluating claims that are presented to you, or at least I hope you will have by the end of the book. Fail to use these tools or assume that you are too smart to be taken in and that is the point you find yourself investing all your money in a multi-level marketing scheme. As any magician will tell you the smarter and more confident you are in your ability not to be fooled the easier it is for you to be taken in by some simple but effective trick.*

---

* Many magic tricks are far simpler than they appear. I actually recommend that you take the time to learn a few easy card tricks and then show them to your friends. You will be surprised by how easily impressed they are with a trick that you, knowing how they are done, are sure they must be able to see through with ease. There are many great card trick tutorials on YouTube. I recommend doing a search for mismag822 as his videos take you through the basics of performing tricks as well as providing easy to follow instruction on how to do numerous different card tricks. I really believe that learning a few card tricks will teach you a more fundamental lesson about skepticism and critical thinking than everything else in this book and as such in the next chapter I will teach you a simple trick that you can do on your friends and family. But do keep reading after that anyway.

Science! It Works, Bitches!

There is a concept in psychology called the fundamental attribution error that basically goes like this. If you are walking down the street and you trip over you will immediately start looking around for the loose paving slab or untied shoe lace that caused you to trip. However, if you are walking along and you see someone else trip over you will problem jump to the assumption that they are clumsy or weren't paying attention to where they were going. It is the exact same event and yet when it happens to us we look for some external factor to blame, but when it happens to others we assume it must be due to some internal flaw that the person possesses. This applies perfectly when dealing with skeptics and believers. If we get taken in by something then we blame the situation, we'd had a long day at work and weren't firing on all cylinders, we weren't presented with all the evidence, the person cheated in some way. And yet when we see others falling victim to some wild claim or idea we instantly assume they must be foolish or unintelligent, when it is more than likely that it was the situation, rather than some personal flaw, that caused them to be taken in. That is worth keeping in mind.

Most people who accept paranormal or pseudoscientific claims fall into the category of being mistaken rather than unintelligent. That said it is completely true that some people do hold stupid beliefs because, well, for the very reason you would imagine someone would hold a stupid belief. I also want to make it clear that I feel that there are times when mockery is a valid tool to have in your skeptical toolkit. Pointing out the absurdity of a belief or idea using humour can be an effective way of getting your point across. I also feel that there is a fine line between using mockery to highlight the flaws in someone's thinking or to try and get them to see things from a different point of view and using it simply to belittle or put them down. Mockery and humour are, to my mind, best used like a scalpel rather than a hammer.

As we go through the rest of the book I may occasionally poke fun at some of the bizarre beliefs out there but I do not want you to think, for the most part anyway, that I am mocking the believers or assuming they are stupid because of the things they believe. I have said it already, and no doubt will repeat again, but a good skeptic should always keep in mind the possibility that they are the one in the wrong and that tomorrow new evidence might turn up to show that this is the case. But, as we shall see, this should never be thought of as a bad thing. Being shown you are wrong is a great thing for a skeptic as it gets us one step closer to the truth. So let's get on with it shall we?

Science! It Works, Bitches!

Skeptical Sidebar #1
The behind the back mentalism trick

As I wrote this book I came across a number of topics and ideas that either only tangentially related to the main chapters or which deserved an entire chapter of their own. I also wanted to have a little bit of fun and take a break from some of the more heavy going aspects of skepticism. As such, you will find these Skeptical Sidebar sections scattered between the main chapters in which I will either focus on a single skeptical issue or idea or teach you a few fun tricks and skills with which to impress your friends. Let's get started with a little magic.

As promised here is a simple card trick with which to amazing and entertain your friends. I love this trick as it never fails to impress and always leaves people wondering how you did it and as such I am a little bit upset to be giving away its secrets. However, it is also the perfect trick for a beginner, as it requires no special skills or sleight of hand what-so-ever. It is incredibly simple to do and as such perfect as a lesson about critical thinking. It is so simple that you will be convinced that people will see straight through it and work out how it was done, but they won't. Wait a damn second; my friends are smart people, you say, so of course they will see through it if it is that simple. Well as I said in the previous chapter

17

Science! It Works, Bitches!

sort of thing has little to do with how smart you are and everything to do with knowing the right information and applying it correctly. As such unless they know magic themselves, or you do something horribly wrong, I assure you that they won't see through it and they will be suitably impressed with your skills. Ok on with the trick.

The Set-up

The reason this trick is so simple to do is because it really requires no special skills on your part as it pretty much performs itself. Everything that makes it work is in the set-up of the trick and not the performance and as such, unless you go terribly wrong, this trick will work for you the very first time without any practice. Ah, if only everything in life was that easy.

Ok the first thing you need to do is sort the pack into suits. Each suit needs to be arranged with the King at the bottom of the pile with the Queen next, then the Jack, then the Ten and so on ending with Ace on top.

Next you need to arrange the suits in order of Hearts, Clubs, Diamonds and Spades. Technically it doesn't matter which order you place them in as long as you go red, black, red, black. However for the sake of argument go with this order for now.

Building your Skeptical Toolkit

Now take the pile of Clubs and search through it until you find the Jack. Then take that card and the cards below it and place them on top of the pile. Now take the Diamonds and find the Eight and place this and all the cards below it on top of the pile. Finally do that same thing with the Spades, only this time locate the Five and move it and all the cards below to the top of the pile.

Now you need to sort the cards together. Start a new pile, starting with the Ace of Hearts and place the Jack of Clubs on top, followed by the Eight of Diamonds and then the Five of Spades. Work through the piles, taking one card at a time and always in the order of Hearts, Clubs, Diamonds and Spades, until you have piled all the cards into one stack. If done correctly the top four cards on the new pile should look like this:[*]

---

[*] For those of you interested in the technical aspects of the trick the cards are now sorted according to something known as the Gilbreath Principle.

Science! It Works, Bitches!

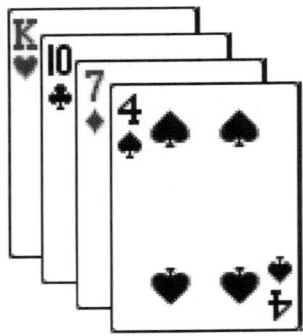

Place all the cards back in the box and head off to find your victims, I mean audience. Personally, and this is not important to the trick, I like to put the Jokers and the Instruction cards on top of the cards in the pack as well. When you take the cards out of the pack you can make a show of removing these and placing them back in the box. This adds to the illusion that you are simply taking the cards from the box as you found them rather than in a prepared state.*

I won't try to explain it here but very simply it is the idea that when cards are sorted in a specific way certain outcomes are mathematically certain. You can find a great explanation of how the Gilbreath Principle works, as well as other card related mathematical conundrums, here - http://www.isid.ac.in/~rbb/lsrarticle.pdf

* Another thing I have found performing this trick, and that again is not actually part of the trick itself, is that if you use an unusual looking pack of cards people will assume that the pack you are using has something to do with how the trick works. You can play on this idea to your benefit by dropping false hints about how you are doing the trick or by encouraging your audience to check the cards you deal for themselves. They won't guess how you did it anyway but a little extra misdirection never hurts.

The Trick

Remove the cards from the box and fan them out slightly so that your audience can see that they are all mixed up then quickly bring them back together and place them face down on the table; we don't want them getting too good a look. Now ask a member of your audience to cut the pack anywhere they want, place the upper half of the pack on the table and then the lower half of the cut on top of this. Get them to do this as many times as they would like. If you are doing the trick to more than one person it might be fun to get them all involved. Get them all to take a turn at cutting the pack in this fashion as many times as they would like. I like to say something like "well unless you are all in on the trick those cards and now pretty well mixed up."

Next get a member of your audience to cut the pack into two equally halves and place one half face up on the table and the other half face down.

Now get one of them to do a riffle shuffle. In case you are not sure this is where you take half of the deck in each hand, holding the two halves with your thumbs inward, then use your thumbs to release the cards a few at a time so that they fall to the table interleaved. Once they have done this get them to hand you the pile and fan it out slightly so that they can see the cards are all mixed up, but again not for so long that they can get an idea of the order. You should end up with it looking something like this:

Place the cards behind your back, with the top of the pack facing upwards. Now comes the magic. Though the actually trick performs itself there is still an element of showmanship to this. You may like to invent a story to explain how you are about to do what you are about to do. One story I like to use is to claim that I have trained myself to be able to tell the difference between the cards simply from the feel of the different coloured inks used. This is another reason I like to keep the Jokers in the box as well as it gives you the option, if you feel like doing so, to get your audience to remove them from the box and see if they can tell the difference between them just by touch. They won't be able to because there is no difference, but it all adds to the story you are telling. Go with whatever story you feel comfortable with.

No matter what story you decide to use tell your audience you are going to sort through the pack and, without looking, find a red and a black card. Take a couple of seconds to pretend you are doing this during which time simply count the first two cards off the top of the pack and place them on the table. They *will* be a red and black card. Now your audience might not be all that impressed by this and I like to admit that, well, I did see the colour of the top card when they handing me the pack and I had a

Building your Skeptical Toolkit

fifty/fifty chance of picking a card of the opposite colour at random so, yeah, it's not all that impressive. So you tell them you will do it again.

Again make a small show of sorting through the pack behind your back while you again count off the top two cards of the pack and place them on the table beside the first two. Again, unless something went really wrong in the set-up, they *will* be a red and black card.

Do this two more times, telling your audience that you are doing so to show that it is not just luck, until you end up with four pairs of red and black cards on the table. Ok now it is time to up your game. Tell them that now you are going to sort through the pack and find one card of each suit. I like to throw in comments to make it sound like I am really searching through the deck. Say things like "hmmm, no I think I will go with this one" or "maybe one from the bottom of the desk" or "almost dropped one there", anything that supports the idea that you are searching the deck for the cards you need. Of course once again all you need to do is count off the top four cards to end up with something like this:

Science! It Works, Bitches!

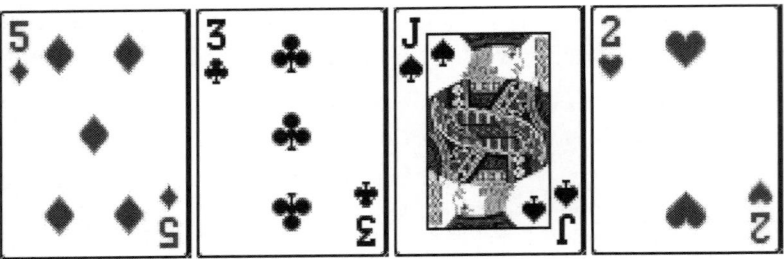

Repeat this three more times so that you end up with four sets of cards each with one card from each suit in it. By now your audience should be suitably impressed so it is time for the big finish. Before we do this, we need to lose two more cards from the deck. The next two cards will be a red and black card so simply repeat the red and black prediction you did earlier. This may seem like you are moving backwards so you may want to give a reason for doing this. I usually say something about feeling that the last of the four suit card deals was more luck than judgement and that I'm just checking that my senses are still with me. Whatever you say you need to burn two cards.

Ok big finish time. Tell your audience that you are now going to attempt to find a run of 13 cards all the way from the Ace to the King. As you have probably guessed this just involves dealing out the next 13 cards from the top of the pack, but remember you are meant to be finding them so take your time with this. Once you

Building your Skeptical Toolkit

have the 13 cards in one hand and the rest of them in the other bring them back around the front of you. Place the remainder of the cards on the table and then work through the 13 cards in your hand and deal them out in order of Ace to King. Again you will find that all the 13 cards you need to do this are there somewhere.

By now your audience should be really impressed, but you are not done yet. Pick up the remaining cards from the table. You will find you have 13 of them and that, yes, once again you can deal them out all the way from the Ace to the King.

Stand back, take a bow and try to resist the urge to tell them how it was done.*

Alternate ending

Now while that is how the trick should be done I often find that people mistakenly assume the trick has finished when you reveal the first set of 13 cards. Now depending upon your performance you may be able to turn this to your advantage, alternatively you might like to change the ending slightly. One thing you might like

---

* That's the part I always fail at.

Science! It Works, Bitches!

to do is get your audience involved in the trick once more. When you bring the two piles of cards out from behind your back place them both on the table in front of you and ask a member of your audience to pick one of the two piles. You might like to jokingly try to get them to pick one pile over the other but at the end of the day it doesn't matter which one they choose. Place the pile they didn't choose back in the box and get them to pick up their chosen pile. Now simply get them to finish off the trick themselves, counting out the cards from Ace to King onto the table. This added element implies that you were somehow able to predict which pile they would choose in addition to finding the right cards.

At the end of the day just have fun with it. If you think of something to add that will make the trick better then do so. As I have said many times your audience will not be able to guess how you did it, even if your performance of it is not all that good. They will most likely assume that you are using a trick deck or some impressive form of sleight of hand, and I say let them. When all is said and done you will have amazed them with a simple trick and will have learnt a valuable lesson in skepticism and critical thinking along the way.

Just the facts ma'am
Facts, hypothesis, theories and laws

There is a common saying amongst some of the more prominent skeptics that pretty much sums up the main goal of skepticism.

> "A good skeptic aims to believe as many true things as possible, and disbelieve as many false things as possible."

This is a good starting point for a beginners guide to skepticism and in this modern, 21st century world if this is something you want to do then you need to have a basic understanding of science and the way it works. As such in the first part of this book I want to look at the methodology behind good science and logical analysis and the reasons why this offers us a far better way of discovering the truth than alternative ways of thinking. Don't worry, I'm not going to bombard you with complex discussions of thermodynamics and quantum physics, though I will mention both due to the prevalence with which purveyors of pseudoscience often misuse both these subjects, instead I will be focusing on a more general view of the structure of scientific inquiry. But before we do that it is a good idea just to take a moment to look at why it is important for everyone, not just the good skeptic, to have an understanding of basic science. As always Carl Sagan[*] put it best:

---

[*] Seriously I cannot emphasis enough the need to read this great man's books. There is a very good reason why he is so revered by both the scientific and skeptical communities. The man was a poet of science and reason.

Science! It Works, Bitches!

> "We've arranged a global civilization in which most crucial elements — transportation, communications, and all other industries; agriculture, medicine, education, entertainment, protecting the environment; and even the key democratic institutions of voting — profoundly depend on science and technology. We have also arranged things so that almost no one understands science and technology. This is a prescription for disaster. We might get away with it for a while, but sooner or later this combustible mixture of ignorance and power is going to blow up in our faces."

We live in a world where it is arguable that every single aspect of our daily lives owes its existence to modern science. The computer I am writing this on is an obvious example, but then so is the chair in which I'm sitting, made as it is from various plastics. The clothes I am wearing are primarily made from artificial fabrics. The food we eat is the result of modern agricultural farming methods, which owe a debt to science, to say nothing of whatever form of transport was used to bring that food to our local store and the methods used to keep it fresh.

Now while it is unrealistic to expect people to become experts in all of the many varied areas of science that make our modern lives so much easier than they would have been just a century ago, an understanding of the way in which science works and the methodology it uses to come up with answers is something that everyone should know. If, as Carl hints at, all of modern science were to disappear tomorrow then the most important piece of information for us to retain would not be any particular finding of science, but rather the methodology that allowed scientists to make those findings in the first place.

But before we get onto that I want to start by looking at some of the basic terminology used by science and why an understanding of this is vital for every wannabe skeptic.

Facts

Scientists have a habit of using words in a way that differs from how they are generally used in normal conversation. The word "fact" is a good example of this. To most people the word "fact" refers to something that is undoubtedly, unquestionably, one hundred percent true. We talk about things that we know for a fact, meaning things that we know for sure are true, and it is generally accepted, and in many cases may often appear, that scientists use the word in the same way. This, if you will excuse the common usage, is not in fact correct.

When a scientist uses the word "fact" they are talking about something that, at that moment and in that situation, is assumed to be true, but which it is entirely possible may be refuted at some point in the future. In science a fact is something that has been objectively verified by way of repeated observation and that has a real, verifiable existence and so is accepted as true. The late, great Stephen Jay Gould, a renowned palaeontologist, evolutionary biologist and science historian, put it like this:

> "Moreover, 'fact' doesn't mean 'absolute certainty'. The final proofs of logic and mathematics flow deductively from stated premises and achieve certainty only because they are NOT about the empirical world. ...In science 'fact' can only mean 'confirmed to such a degree that it would be perverse to withhold provisional consent.' I suppose that apples might start to rise tomorrow, but the possibility does not merit equal time in physics classrooms."

Gould mentions the example of falling apples as something science would consider a fact. If any of us, at any place on the planet, were to drop an apple it would, without fail, fall to the ground. Any number of people can repeat this simple experiment any number times and get the same result. As such the falling apple has met the first criteria for being a scientific fact, in that it has been "objectively verified by way of repeated observation." As for the second part, an apple is an object that can be measured and quantified and which has a tangible presence and so can be said to have a "real, verifiable existence," thus meeting the second criteria for a scientific fact.

Now even though in every single experiment conducted with a dropped apple the apple falls to the ground and as such is classified as a "fact" this does not mean that science rules out the possibility, no matter how incredibly small, that this will not always be the case. With this caveat in place scientific facts form the foundation of all science. Because of the way "facts" are defined by science it is generally accepted that they will remain the same regardless of who is observing them, no matter who performs the experiment involving them, and will always produce the same outcome under the same conditions.

Before we move on it is worth taking a moment to look at the difference between facts and beliefs. It is not at all uncommon for someone to describe a belief they have as a "fact," and while it is possible that they may be completely right this does not mean that their belief meets the standard required to be classified as a scientific fact. The belief, for example, that there is life after death may ultimately turn out to be true; however it cannot be classed as a scientific fact. For a start there is no way that such a claim can be "objectively verified by way of repeated observation." We can't kill people off and then ask them to report on their surroundings, and even if we did this and brought the person back and asked for a report it would be highly subjective, coloured by the persons existing beliefs and by the fact that any claim they make about observations they made would be highly suspect given that their brains were dead at the time. We could not objectively verify anything they said.

The same holds true for the second part of our scientific fact equation. Life after death is not something we can measure or quantify and does not have any kind of tangible presence. As such we cannot class it as having a "real, verifiable existence." That's strike two and in this game that means you're out.

Hypothesis

A hypothesis is an educated explanation based upon one or more observable phenomenon. In order to be a scientific hypothesis it must be possible to test the proposed explanation to see if it stands up. It must also be presented in the form of an "if-then" statement rather than as a question. Let me give you a real life example.

Back in the first century BC the Greek philosopher Aristotle, drawing on the work of many others, proposed an idea that would become known as the theory of Spontaneous Generation. Spontaneous Generation is basically the idea that some forms of life arise from sources other than parent organisms, seeds, eggs etc. An example of this is the idea that maggots arose from rotten meat. The theory of spontaneous generation was ultimately disproven by Louis Pasteur in the 19th century; but let's imagine that it wasn't. How would we go about investigating this claim?

As with all scientific hypotheses we start with an observable phenomenon. If a plate of rotten meat is left outside maggots eventually appear. This is something that can be observed by

Science! It Works, Bitches!

pretty much anyone and which, using the definition defined above, would count as a basic scientific fact. Now let's say that, like Aristotle, we believe that this is because the maggots are spontaneously generated from the rotten meat rather than being caused by any outside agent, how would we go about testing this? Well the first thing to do is put our hypothesis in the form of an "if-then" statement.

> IF maggots are spontaneously generated by rotten meat rather than being caused by some outside agent, THEN we should be able to cover the meat and get the same result.

This is an easy experiment and one that, should you be so inclined, you could carry out yourself. If maggots are spontaneously generated by the rotten meat itself then covering the meat up to prevent access to it from an outside agent, such as flies, should have no effect on whether you get maggots of not. This is a good example of a scientific hypothesis and in fact is exactly what Pasteur did to show that the theory of Spontaneous Generation was wrong.

The most important part however comes when you get the results of your experiment. If your experiment shows your hypothesis to be false then it must be abandoned or modified. On the other hand if the experiment produces results that fit with your hypothesis then it can be used as a basis for further investigation and for developing more complex inferences and explanations.

Theories

Ok, this is the big one. Most people generally don't understand how the word "theory" is used by science as, once again, those pesky scientist use it in a rather different way to how it is used in general conversation. Have you ever heard someone say something like "you can't trust that, it's only a theory" or "that's just a theory, it's not a fact"? I imagine that you probably have and this is because, when part of everyday usage, "theory" pretty much means "guess" or "hunch." In fact, most of us use the word in a similar way to how scientists use the word "hypothesis."

Detectives look at the evidence and have a "theory" about who committed the crime. You come home to find a mess in your kitchen and have a "theory" as to how it happened. We use "theory" to describe a proposed explanation whose validity may still be in question and that has not yet been established as "fact." And in both cases we use those words in a very different way to how scientists use them.

In science "theory" most definitely does not mean "guess." In science the word "theory" is used to describe a hypothesis or set of hypotheses that have been extensively verified by empirical data, evidence and repeated testing and that have been generally accepted by the majority of scientists working in the field as a valid way of explaining an observed phenomenon. Scientific theories are an attempt to understand and explain one or more natural observations.

It is important to understand that a scientific theory never becomes a fact. Scientific theories explain facts. One of the most commonly misunderstood examples of this is the theory of evolution. At its most basic evolution is the observed fact that the allele frequency of genes in a population changes over time. An allele can be thought of as a variation of a given gene. Let's say

Science! It Works, Bitches!

that there was just one gene that controlled eye colour. Now this gene could have various alleles, one for brown eyes, one for green and another for blue. If the number of people with blue eyes in a given population changes over time then this is evolution in action. I know, most people only think about evolution on the grand scale of dinosaurs turning into birds, but this kind of small scale change is just as much evolution as the big stuff. That evolution happens, that the allele frequencies of genes do change over time within populations, is a fact. The theory of evolution is currently the best explanation for how and why this happens. As such in this case evolution is both fact and theory, using the correct scientific definitions of those words.

On top of this a good scientific theory needs to have predictive power. As with hypotheses, which can pretty much be thought of as unaccepted theories, a scientific theory needs to be able to make "if-then" statements that can be tested and verified to see if they are accurate or not. Once again, as with hypotheses, if a theory is shown to be false then it must be discarded or modified to fit with the evidence. Evidence is never made to fit the theory; it is always the other way around.

Finally a scientific theory needs to be falsifiable. All this really means is that there should be a good understanding of what it would take to disprove the theory. If you are ever presented with a "theory" for which there is no way to disprove it then it is not a valid scientific theory. The classic example often used to demonstrate this idea brings us once more back to the topic of evolution. So what would disprove the theory of evolution? Well how about finding a rabbit fossil, or any mammal for that matter, in Precambrian rock.[*] This would go against everything that

---

[*] Ok technically this would not disprove the theory of evolution, that being the idea that changes in the allele frequency of genes are the result of random genetic mutation and natural selection, as there is simply too

evolutionary theory tells us about the way in which life on this planet developed and as such making such a find could potentially falsify the theory.

Laws

Probably the first and most important thing for you to know about scientific laws is that, as with facts turning into theories, theories never become laws. A scientific law is a description of something that is accepted as a universal and invariable fact about the physical world. Unlike theories and hypotheses scientific laws do not attempt to explain anything, rather they are analytic statements about the universe that often, though not always, may be presented in the form of an equation or contain an empirically determined constant. Easily the best example of this is the one bit of advanced physics that everyone knows, and that's Albert Einstein's law of mass–energy equivalence.

$$E = MC^2 \text{ (Energy = mass} \times \text{ the speed of light squared)}$$

This simple equation describes the relationship between energy and mass and contains the empirically determined constant of the speed of light. Scientific theories may contain many such laws or may be implied based upon verified laws, however there is never a point when a theory transforms into a law. It is a sure bet that if you ever hear someone saying something along the lines of *"it's only a theory, it's not a law"* then you are dealing with someone

---

much supporting evidence for that. However, it would call into serious question the established order of evolutionary development as mammals did not arise until over 300 million years after the end of the Precambrian period.

Science! It Works, Bitches!

who does not understand the scientific meaning of either of those terms.

And again, as with everything we have talked about so far, a scientific law is never viewed as the last, indisputable word on any issue. If new evidence comes up that disproves the law then, as with everything else in science, it must be abandoned or modified. Also, laws may not apply in all situations[*] or may only describe things to a certain level of detail. For example, Sir Isaac Newton's laws of gravity may not explain things as accurately as Einstein's theory of general relativity and yet they were more than sufficient to put a man on the moon.[†]

Before I finish this chapter there is one more word I want to touch on that you may have noticed I purposefully didn't use in any of my descriptions. That word is "proof." Science doesn't deal with proof. No matter how much evidence you have, no matter how many tests you run and no matter how closely the results match your predictions nothing in science can ever be described as proven. Now the public and the media love to talk about the scientific proof for this or that claim, when what they are actually talking about is evidence.

The reason you can never prove something in science is because you can never be sure you have accounted for every possible situation, variable or alternative. True, apples have always fallen down in every single experiment we have ever done and as such we can give our "provisional consent" to the idea that they always do so, but does that mean that we can completely rule out the possibility that somewhere, in some situation that this may not be the case?

---

[*] For example, the 2nd law of thermodynamics only applies to closed systems.
[†] The distances and speeds involved did not require the use of more precise relativistic calculations.

Science can never be proven, only disproved. As such if someone tells you they have proof of some claim or another then they are either using the word in a colloquial fashion or just not talking about science.

Science! It Works, Bitches!

Method in my madness
A quick look at the scientific method

> *"It does not matter who you are, or how smart you are, or what title you have, or how many of you there are, and certainly not how many papers your side has published, if your prediction is wrong then your hypothesis is wrong. Period."*
>
> Richard Feynman, renowned physicist, science populariser and bongo player.

What distinguishes a good skeptic from the crowd is a willingness to abandon even their most cherished beliefs in the face of solid, contradictory evidence. While this may at times be difficult, a skeptic can do this because the ideas they hold and the conclusions they have come to are not as important as the method they used to reach them. Because of this, when new information is presented that shows an existing belief to be false, abandoning that belief is entirely consistent with the method of evaluating evidence used to form that belief in the first place. Nothing is actually lost by throwing out the false belief and this actually helps the skeptic get closer to their goal of believing as many true things and as few false things as they can. For a good skeptic getting rid of bad ideas is always a win win situation.

The methodology that each individual skeptic uses to try and reach this goal can of course vary from person to person. However, there are a number of things that most skeptics seem to have in common. These commonalities generally include an everyday use of the scientific method, the application of the laws of logic and an evidence based approach to answering questions. Over the next few chapters we are going to take a quick look at each of these areas in turn.

The Scientific Method

The Scientific Method is something of a blanket term that describes a number of different techniques used to investigate natural phenomena, acquire new knowledge and correct, verify and, if needed, modify existing knowledge. Now while the specific techniques tend to vary from field to field, in order to be considered science they all have to be based, at their core, on observable, empirical and measurable evidence. All scientific methodology revolves around collecting this evidence and then making, and more importantly testing, hypotheses based upon that evidence. Once your hypothesis has passed this first stage it faces a second major challenge in the shape of peer review, but we will come onto that in a bit. First, let's look at the basic steps involved in pretty much all versions of the scientific method. Now as I mentioned the scientific method has something of an amorphous quality to it and so can be described in a number of different ways. Here I have chosen to break it down into seven basic steps but this does not mean that the scientific method is a strict seven-step process. You may have come across descriptions of it that break it up into more or fewer steps, or even lay it out in a way that seems to include no distinct steps at all. At the end of the day though as long as the description follows the general structure of observation – hypothesis – experimentation – evaluation, then that description is probably on the right track.

Step One - Ask a Question

All science starts with asking a question. Without questions there simply wouldn't be science. These questions are based upon the observation of a phenomena. One of the best and simplest examples is the one that Newton asked. Why did that apple fall from the tree? This simple question set Newton on his course to defining the laws of gravity.

Step Two - Do Research and Gather Evidence

Science is an accumulative process. Whenever we generate a new question we do not have to start completely from scratch with regards to answering it. Instead we can build upon the work of others who have already investigated this question or a related issue. We can learn from them, both by looking at the evidence they discovered and the conclusions they drew from it, as well as the methodology they used to do so. That way we can learn from their achievement as well as their mistakes. Newton himself summed this up perfectly:

> "If I have seen further it is only by standing on the shoulders of giants."

As well as looking at the work of others you must gather your own evidence. Let's use Newton's apple as an example. It is possible that the apple Newton saw fall from the tree was special in some way and so this single data point is not enough to start forming any kind of hypothesis just yet. You need to gather more data. Was what happened to this apple a one off or does it happen to

Science! It Works, Bitches!

all apples? Do other objects act in the same way? Observe and record and only when you have an adequate body of evidence can you proceed to the next step.

Step Three - Construct a Hypothesis

This is the point where, based upon the evidence you have gathered and the research you have conducted, you attempt to come up with an answer to the question that started you on this process in the first place. Why do apples fall from trees? Maybe there is some kind of physical force pushing them down? Once you have your hypothesis you can move onto the next step, which is:

Step Four - Make a Prediction

This step and the previous one are very closely linked and as such are often explained as a single point. I decided to divide them up to make things a bit clearer. A prediction is something that you think will happen under certain circumstances if your hypothesis is true. Now I purposefully made the hypothesis above incorrect in order to better demonstrate the next few steps. Based upon this hypothesis we might make the following prediction:

> IF I place something over the top of the apple THEN it should no longer fall.

Ok so that is a terrible example but it makes my point. Predictions usually follow the formula of "If I do X then Y will happen." Once you have your prediction it is time to test it.

Step Five - Perform an Experiment to Test Your Prediction

An experiment allows you to test whether your prediction, and by extension your hypothesis, is true or false. A good experiment will only test one variable at a time, making sure that all other conditions stay the same and so allowing you to be fairly confident that the thing you are testing is causing the results to change, or not. You should also repeat the experiment a number of times to make sure that the results are consistent. It's possible that you did something wrong the first time without noticing and so repeating the test a number of times will reduce the chances of error.

Step Six - Analyse Your Data and Draw a Conclusion

This is the most important part. Now that you have the results of your experiment you need to analyse them to see if they match your prediction and thus support your hypothesis. If they do then great, it seems you may be onto something. However if they do not and your hypothesis is wrong then it is what happens next that makes science so powerful, though there are those who may see it as a weakness. You go back to the beginning, form a new hypothesis and start again. You don't stick with your failed hypothesis in the face of evidence against it. No matter how much you may like a particular hypothesis, if the evidence does not support it then it must be rejected. Actually, even when a hypothesis is shown to be correct good scientists will go back, come up with a new prediction and experiment and test it again to make sure. Only then will they move on to the final stage.

Step Seven - Share Your Results

Once you have the evidence and have analysed it and found that it supports your hypothesis, it is time to tell the rest of the World about it. Now in our modern world it would be very easy just to go to the media or stick your results up on the internet, and some people do indeed do this. However, good science never takes this straight to the general public approach and instead goes through a process known as Peer Review. The Peer Review process is basically the gateway to publication. You submit your hypothesis and the evidence supporting it to a journal run by recognised experts in your field. They will then go through and check your article for errors, making sure you correctly followed the scientific method and that the evidence presented really does say what you think it says and that it fits with other known evidence within the field. Most times your first attempt to get published will be rejected and the experts will come back with questions and issues they want you to address before they will publish your work. Once they are happy that you have correctly followed the scientific method and that the evidence accurately reflects your conclusions then your work will be published in a scientific journal that specialises in that field and shared with the World at large.

So there you go, the scientific method. But wait; there is still one more step to go. Now I am sure that those of you with sharp minds will have noticed a potential flaw in this process. Let's say that I'm less than honest or just lazy. I come up with a hypothesis, conduct an experiment and find that the results do not support that hypothesis. Now I should go back to the drawing board and start again, but I really liked that hypothesis and I really don't want to do all the work again. And so I fake my results to make it appear that the evidence did indeed support my hypothesis. I submit it to the journals and because everything appears to be in order, the evidence looks like it supports the hypothesis and it appears I have followed the correct methodology my work is published. Now this is rare but it does happen, so does that mean that the scientific method fails?

Well no, and that's because of what happens next. Other scientists come along, look at my evidence and try to replicate it. They repeat the experiments that I did and look to see if they come up with the same results. If they don't then it will indicate that there is something wrong somewhere. They repeat the tests and get other scientists in other labs to do the same thing. Still they can't get the results I did. At this point maybe they come up with their own hypothesis as to what is really going on and test this and this time the results do fit with the predictions. These scientists then write up their results showing why I was wrong and submit them to peer review and ultimately get them published. My fake results start to look more and more suspect and as the evidence mounts up they are at first marginalized and then completely dismissed.

The perfect example of why following the peer review process is important and equally of how science is self-correcting is the famous Pons and Fleischmann cold fusion experiment. To cut a long story short, in 1989 scientists Stanley Pons and Martin Fleischmann believed they had made a breakthrough in the area of cold fusion, nuclear fusion at low temperatures. Instead of going through the peer review process they announced their findings directly to the media and the general public. The news rocketed around the World and the media heralded it as the dawning of a new age. But when other scientists tried to replicate Pons and Fleischmann's results they were unable to do so. As more and more scientists reviewed their work it quickly became clear that something was very wrong with their claims. Ultimately it became obvious that they were fundamentally wrong in their conclusions and that cold fusion, and the new age, would not be happening any time soon.

This sort of error correcting is exactly the sort of thing the peer review process is designed to do. It is only because Pons and Fleischmann bypassed the process that the general public became aware of their failure. Had they gone through the peer review process then the end result, their work being discredited, would have been the same, but it is likely only people working in that specific field would have ever been aware of it.

Science can be harsh but there can be great benefits, both to the World as a whole and to the specific scientist in question, if one scientist is able to show that another wrong. As such scientists all over the World are constantly testing and re-examining things that many of us just take for granted. There is something of a "fastest gun in the west" mentality in science. Young scientists are desperate to prove the big names wrong and thus gain fame and accolades for themselves. Coming back to a favourite subject of mine for a moment, one of the most common responses to critics who argue that evolution is not supported by the evidence, and only put forward as valid because of some secret conspiracy of scientists, is to point out that the fastest way to a guaranteed Nobel prize is to demonstrate that evolution is wrong. Though science can be stubborn and set in its ways at times it is, at its heart, a search for truth. Any scientist who could show, with good observable evidence and valid repeated experimentation, that some part of evolutionary theory was wrong would not be dismissed and silenced, but rather would quickly find their name joining the ranks of scientific greats alongside the likes of Newton, Einstein and Darwin himself.

A law unto themselves
The three laws of logic

As well as a good understanding of the scientific method, by which evidence can be evaluated and sound conclusions reached, a good skeptic should be well acquainted with the three founding principles of informal logic, the form of logic generally used in debates and arguments.

Whenever you find yourself in discussion with a believer or proponent of some pseudoscientific or paranormal claim an understanding of basic logic is important for you to effectively be able to get across your reasons for being skeptical, as well as being able to identify flaws and fallacies in your opponent's arguments. In the next chapter we will take a look at some of the more common logical fallacies you are likely to come across, but before we do that it is a good idea to quickly go over the three "laws" of informal logic, how they work and why they are important.

Before we do I have a confession to make. Though I have always thought of myself as someone with a scientific outlook on life it wasn't until recently, when I first became aware of skepticism, that I learnt about the laws of logic. Almost as soon as I did I found myself thinking back over conversations that I'd had and I was fairly quickly able to identify where the arguments that I had presented, and at times had even won the debate with, failed in their logic and as such were in fact poor and flawed arguments. By understanding the laws of logic and making sure that any arguments you present adheres to them you can only strengthen your side of the discussion.

Science! It Works, Bitches!

The Law of Identity

As Popeye the Sailor Man was fond of pointing out, "I yam what I yam," and this simple idea is pretty much all there is to the law of identity. This principle of informal logic states that something has an identify, meaning it is what it is and conversely it is not what it is not. For example a dog is a dog and it is also, if you will excuse the double negative, not not a dog. This is one of those things that people tend to take for granted, but it is in fact a fundamental component of any debate, or any conversation for that matter. If you and your opponent are unable to agree that a dog, or whatever it is you are discussing, is indeed a dog then you will never get anywhere. Generally, we have no problem with the idea that a dog is a dog, a book is a book and a car is not not a car. This is actually a pretty good thing or we would all have a lot more difficulty getting anything done that involved more than one person.

The Law of Non-Contradiction

This principle holds that something can't hold two contradictory properties at the same time and still be valid. A fire cannot be both hot and not hot at the same time. This is what is known as a conjunctive proposition, that being a proposition that uses the word "and", as in "this and that." In the example above we have the proposition that "the fire is hot AND the fire is not hot." As these things are contradictions then the proposition itself must be false.

It is worth noting that in the above example it is simply taken as read that I am saying that the fire is hot and the fire is not hot at the same time and in the same place. In everyday use we don't tend to explicitly state these details and assume people will simply fill in the gaps. This is worth remembering, as something does not violate the law of non-contradiction when what you are describing takes place at different times and places. Let's swap from fire to water to better explain this. Now if I say the water is hot and the water is not hot at the same time and in the same place, well then I have fallen foul to the law of non-contradiction. However, it is completely possible for the water to be hot and the water to not be hot if we are talking about different places and/or times for the two variations. It is also possible for the water to be partly hot and partly cold as the same time, though we would probably just call it warm. These points should be kept in mind during a debate when addressing something that appears to violate this law. Always make sure that your opponent is indeed talking about two contradictory properties existing at the same time and place before pointing out the flaw in their logic.

The Law of Excluded Middle

The final law of informal logic applies to propositions that make a related disjunctive claim, or in other words sentences that have the word OR in the middle. This applies when you have two statements and one of them is true or the other one is. A good example of this is one that Brits like me are very familiar with, it is raining OR it is not raining. One or other of these two statements must be true. Now we already know from the law of non-contradiction that they both can't be true at the same time and place so why have this law at all? Well the law of excluded middle refers to the non-existence of a third option. It is either raining or it is not; there is no third choice, or to put it another way we have excluded the middle option. A very good way of remembering this one is the saying "you can't be a little bit pregnant." Being pregnant is a situation where you either are pregnant or you are not, there is no middle ground.

In order to make an argument that is logically valid it must adhere to these three laws. However it is worth noting that just because an argument is logically valid that does not necessarily mean that it accurately reflects reality or that it is true. For example, a sound logical argument could be made for a flat earth and yet the evidence would still show that the earth is a sphere. Logic is important and is a powerful tool for use in verbal arguments but all good science should always come back to the evidence. And if something is supported by good evidence then the logic should flow naturally from there.

Skeptical Sidebar #2
Syllogistic Reasoning

Having talked about the laws of logic let's take a quick look at one form of logical reasoning. This will also help expand on the idea that validity and truth are not always the same thing when it comes to logic that I touched on at the end of the previous chapter. Ok so syllogistic reasoning is a form of logical reasoning based upon, unsurprisingly, syllogisms, which are arguments made up of several parts. Typically a syllogism is made up of a couple of statements, or premises, that are taken to be true and a conclusion based upon these premises. A syllogism starts with a Major Premise which is a general statement that as stated is taken to be true, at least within the boundaries of the argument. Next you have the Minor Premise which is a more specific statement that relates to the major premise and again is always treated as being a true statement. For the conclusion, referred to as a putative conclusion within a syllogism, to be logically valid it must be guaranteed by the premises. It is not enough for the premises to simply suggest the conclusion, it must 100% guarantee it to be the case. Let me give you a classic example.

Science! It Works, Bitches!

Major Premise	All men are mortal
Minor Premise	Socrates is a man
Putative Conclusion	Socrates is mortal

In this example the conclusion is guaranteed by the two premises. If all men are mortal and Socrates is a man then it is logical to conclude that Socrates is mortal. Let's look at another example.

Major Premise	All men are animals
Minor Premise	Some animals can swim
Putative Conclusion	Some men can swim

Now at first glance this may seem perfectly fine. I mean we know that men are animals and we also know that some animals can swim and we also know that some men can swim, heck if you are a man reading this there is a very good chance that you yourself can swim. So again this is valid logic...right? Well no and it's because the conclusion is not guaranteed by the premises. Let me change this example a little so that the logical flaw is a bit more apparent.

Major Premise	All men are animals
Minor Premise	Some animals can fly
Putative Conclusion	Some men can fly

Once again we know that our two premises are true, all men are animals and yes some animals, such as bats and birds, can indeed fly. However, this time it should be easier to see that the conclusion does not follow from the premises, because we know that men cannot in fact fly. In these two examples we can start to see how validity and truth are not the same thing. In the first example it is true that some men can swim, however the logic of this argument is not valid. In the second one the conclusion is both invalid and untrue. And, as I am sure you have guessed, it is possible to have a conclusion that is valid and yet untrue at the same time. For example:

Major Premise	If unicorns exist then dragons exist
Minor Premise	Unicorns exist
Putative Conclusion	Therefore dragons exist

The logic of this example is perfectly valid, the conclusion is 100% guaranteed by the premises, however sadly in the real world neither the premises nor the conclusion are true. This is where a lot of people fall down. They assume that if the conclusion is true in the real world, i.e. some men can actually swim, then it must therefore be logically valid as well; and vice versa if the conclusion is not true in the real world, alas dragons do not exist, then they assume the logic must be invalid. This is not the case. Syllogistic arguments should be treated as their own separate little universes in which all that matters is whether the argument is logically valid. Everything discussed in these little universes is treated as true, even if this truth is not reflected in the real world. Now some of you may be scratching your head after that and trying to work out why therefore the some men can fly example is invalid. I mean if an argument does not have to be true in the real world in order to be valid then it doesn't matter than men can't really fly, all that matters is the logic. Well you would be completely correct, the fact that men can't really fly does not matter at all. However, the conclusion is still logically invalid. Let me show you why with a Venn diagram, oh the excitement.

Science! It Works, Bitches!

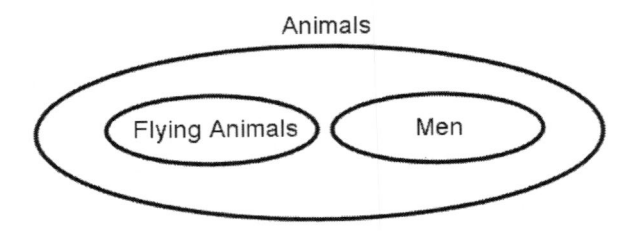

[Venn diagram: "Animals" oval containing two separate inner ovals labeled "Flying Animals" and "Men"]

As you can see from this graphical representation of the syllogistic argument just because "Men" fall within the "Animals" bubble does not mean that they also fall within the "Flying Animals" bubble as well. These two things remain separate and thus the conclusion that some men can fly is logically invalid, as well as being untrue. Now all of that can be a bit tricky to follow at first, and believe me I have only scratched the surface of syllogistic reasoning, so let me leave you with another example. Here we have a line of reasoning that is a little closer to what you might encounter as a skeptic. Keeping in mind what we have learnt about truth and validity see if you can work out if the following syllogism is logical valid or not. I'll even throw in a Venn diagram to help you.

Major Premise	All homeopathic remedies are alternative medical modalities
Minor Premise	Some alternative medical modalities are effective treatments
Putative Conclusion	Therefore some homeopathic remedies are effective treatments

[Venn diagram: two overlapping ovals; left oval contains a smaller circle labeled "X", with "Y" between them; overlap region labeled "A"; right oval contains "Z"]

That's highly illogical Captain
An incomplete guide to logical fallacies

So now that we have looked at some of the basic principles of logic[*] it is time to step over to the dark side and investigate logical fallacies. A logical fallacy is an error or flaw in a logical argument that causes the logic to become invalid. The best way to understand this is to look at them in light of how logical arguments should work. All arguments or debates contain the same basic structure: A therefore B. The person arguing starts by putting forward a premise or two (A), which is usually the fact or assumption upon which their argument is based. They then use a logical principle (therefore) to get to their conclusion (B). A good example of this is a logical principle known as equivalence. Here you would start with the premise that A=B and B=C. From this you can apply the logical principle of equivalence to arrive at the conclusion that therefore A=C. This is a logical argument. A logical fallacy on the other hand is a false or incorrect logical principle and as such any argument based upon a logical fallacy is not valid. Man I hope that made sense.

---

[*] Ok so here is the answer to the question from the previous chapter. The homeopathy syllogistic argument is not logically valid. In the Venn diagram X represents homeopathic remedies, Y represents alternative medial modalities, Z represents effective treatments and A represents alternative medial modalities that are also effective treatments. The premises are that homeopathic remedies (X) are all alternative medical modalities (Y) and that some alternative medical modalities are effective treatments (A). However, this does not mean that some homeopathic remedies are therefore effective treatments, or in other words it does not mean that some X's are A's. So, did you get it right?

Science! It Works, Bitches!

It is important to note that just because a conclusion is reached via a logical fallacy it doesn't necessarily mean that the conclusion itself is wrong; just that the specific argument used to get there is a bad one. In fact there is a logical fallacy, the fallacy fallacy, which actually deals with the idea that just because the argument contains a fallacy the conclusion must therefore be wrong. This is not always the case. Saying that England is a country because grass is green is a fallacious and illogical argument; however the conclusion that England is a country is still correct. Ok, so let's take a look at some of the more common logical fallacies you should keep an eye out for.

The Straw Man argument

The Straw Man fallacy is, to my mind at least, one of the most underhanded forms of illogical argument that someone can use. It is also one of the most common logical fallacies you are likely to encounter and fairly easy to spot. The name of this fallacy is believed to come from a time when military training dummies tended to be constructed from straw. Whilst these dummies looked superficially like a real enemy they were very easy to put down, straw not really being known for its ability to fight back. A straw man argument works in much the same way. The person setting up the straw man will describe a position that sounds superficially like their opponent's actual position but which is much easier to put down. They then attack this argument, rather than the position their opponent actually holds, and claim victory when they are able to defeat it. An example might go like this:

> Barry: I think that the laws related to marijuana should be relaxed.

Building your Skeptical Toolkit

> Steve: But if we allow unrestricted access to drugs then there will be chaos.

In this example Steve may well be right that unrestricted access to drugs would be a bad thing, however this is not the argument that Barry is making. Rather than addressing the actual issue at hand, whether the laws related to the specific drug marijuana should be relaxed, Steve has created a straw man argument about unrestricted access to all drugs that is far easier for him to argue against and much harder for Barry to defend.

Post-hoc ergo propter hoc

If I have a favourite logical fallacy then it has to be this one, and no there is nothing weird about having a favourite logical fallacy. Post-hoc ergo propter hoc is Latin for "after this, therefore because of this" and is the idea that because event two happened after event one then event two must have been caused by event one. This fallacy is at the heart of the "correlation equals causation" problem that we will look at in a few chapters time. Those of you who, like me, have worked in any kind of service industry, and especially if you work in IT, will probably recognise this one. I long ago lost count of the number of calls that I got from people who used this form of invalid logic.

> "One of your IT guys came out and changed the toner in our printer and now my monitor isn't working, what have they done to it?"

> "I never had a problem sending emails until our new temp started. I can't help but think that they are causing the problem somehow."

Science! It Works, Bitches!

Because event one, the IT guy coming out the change the printer toner, happened before event two, the guy's monitor ceasing to work, he comes to the mistaken conclusion that these two events must be casually related, the first causing the second, when in fact they are entirely separate events.

Ad hominem

An Ad hominem argument, or "an argument against the man", is where you target a failing or flaw in the person making the argument and claim that this means that their argument is wrong as well, or alternatively that just because the person has some favourable trait that their argument is also favourable. Here is an example of how this might work:

> "We should not accept Steve's argument in support of allowing gay marriage, he has just got divorced himself."

Unable to defeat the argument directly, in this case an argument in favour of gay marriage, the focus is shifted onto the person making the argument, specifically Steve's recent divorce, or alternatively onto some person or group that can be loosely connected to it and criticizes them rather than the argument directly. The most common form of the Ad hominem argument is to simply insult the person making the argument and claim that this insult invalidates their argument. Claiming, for example, that a politician has bad BO and thus his plan for cleaning up the environment must be wrong or that just because someone doesn't have a job their argument for the existence of Big Foot is wrong by default are ad hominem arguments. That said however it is important to note that while most examples of ad hominem arguments do include some kind of insult it does not follow that just because an argument includes an insult it automatically becomes an ad hominem. For example:

> "Your mother was a hamster and your father smelt of elderberries and thus your argument is wrong."

That would be an ad hominem because the whole argument is based upon the insult of the person. However if you were to say something like:

> "You empty headed animal food trough wiper, your argument is wrong because it does not fit the evidence."

That is not an ad hominem as, while it includes an insult, the insult is not the basis of the argument. So basically you can insult someone all you like and still remain logically valid as long as you don't use those insults as the basis of your argument, which is awesome. Another form of ad hominem argument plays on the idea of guilt by association. An example might go like this.

> "Barry's suggestion to increase the number of buses available to help ease traffic congestion is interesting. But do you know who else was in favour of mass transportation? Hitler."

Here no attempt is made to address the actual points of Barry's bus service argument, instead the whole thing is tarred and feathered by relating it to something with negative connotations, in this case Hitler and his use of mass transportation during World War 2.
This brings us nicely onto the next fallacy.

Science! It Works, Bitches!

Poisoning the Well

This form of fallacy involves making a pre-emptive strike to discredit your opponent, and therefore their argument, by saying something that will produce a biased response in your audience. This is a form of ad hominem fallacy as it tends to be directed at the person as a way of making their arguments seem less valid or trustworthy.

> "Before we look at the evidence I would like to remind people that my opponent was once investigated on suspicion of having committed fraud."

> "My policy will provide the best options for our children, only someone who hates children would disagree with it."

> "And now former Playboy Playmate and famous nose picker Jenny McCarthy will talk to us about the dangers of vaccination."

Before a single point is raised or piece of evidence presented the audience is already encouraged to disregard anything the fallacy maker's opponent is likely to say due to some negative aspect of themselves or the group they represent. Another more subtle, and less intentional, form of this fallacy that you will probably come across might go something like this:

> "Oh that is my favourite film, what did you think of it?"

By saying that it is their favourite film they have biased the person they are talking to into giving a positive, or at least less negative, response in order to not hurt their feelings. Your stated opinion of the film is therefore much more likely to be based on not wanting to upset your friend than on the merits of the movie itself.

Argument from ignorance and the Argument from personal incredulity

These are technically two separate fallacies and yet they are similar enough and stem from the same core mistake to be listed as one. An argument from ignorance is where you claim that a premise is true because it has not been proven false or alternatively false because it has not been shown to be true. A common example of this in relation to science is to claim that a specific field of scientific investigation, such as big bang cosmology or abiogenesis, is false because scientists do not currently understand every aspect of it. Alternatively a proponent of the paranormal may claim that ghosts are real due to the fact the science has not proven them false. This later example often also results in the proponent wrongly trying to shift the burden of proof onto the skeptic.

An argument from personal incredulity on the other hand is, as the name implies, a more personal claim that something is false, or true, because you personally can't think of an alternative. An example of this fallacy is often employed by UFO proponents.

> "It didn't look like any airplane I've ever seen so it must have been an alien spaceship."

> "I've no idea what it was I saw, but clearly they were using some form of trans-dimensional technology."

In both these cases the person is using their own lack of knowledge about what they have actually seen in order to make

Science! It Works, Bitches!

their conclusion. Now it is important to remember that with this fallacy you are not saying that the person committing the fallacy is ignorant in an insulting way. You are not implying that they are a fool or uneducated, but rather that they are using their lack of knowledge about the subject as the basis of their argument. We are all ignorant about something, it is not an insult to have this pointed out, it is how we learn after all.

Special Pleading and Ad-Hoc Reasoning

Special Pleading is a form of fallacious argument that involves someone claiming that something (be it God, ESP, ghosts etc) is exempt from a generally accepted rule or principle but without producing any evidence or logical reason as to why this should be the case. A very common example of this is comes from proponents of pseudoscientific or paranormal claims who state that their claim is somehow beyond the ability of science to accurately test. This means that when these claims are tested, and the results come back overwhelmingly negative, they can turn round and still claim to be correct and that it was the inability of science to accurately test their claim that produced the negative result, rather than their claim simply being false. It may also be suggested that ESP, for example, does not work in the presence of skeptics, which is why all skeptical investigations into such claims come back negative. This is a perfect example of special pleading.
It is also Ad-Hoc reasoning, which basically means making up a reason for something after the fact. This is something that happens a lot when testing pseudoscientific and paranormal claims. A practitioner might claim to get accurate results nine times out of ten and yet when tested perform no better than would be expected by chance. Rather than admit that maybe they don't have special powers they will usually invent an ad-hoc

reason for why it didn't work this time. The spirits don't like being tested, their power doesn't work if people who don't believe in it are nearby, they were not feeling very well at the time, they can't use their power unless it is to directly help someone. Anything but admit the truth that their powers are non-existent.

False Dichotomy

A False Dichotomy is where you reduce all possible options down to just two and then state that if you can show one option to be wrong then the other must be correct by default. This fallacy is often used as a way of shifting the focus away from the fact that you don't actually have any evidence to support your side of the argument by exposing apparent weaknesses in your opponent's argument and claiming these as evidence in your favour. It is known as a false dichotomy or false dilemma because it is never demonstrated that the two options being presented are the only ones possible.

Let's say we have two suspects in a murder investigation, Adam and Bob. We look at the evidence and see that it is not possible for Adam to have committed the crime and therefore conclude that this means the killer must have been Bob. This would be a false dichotomy, showing that Adam is not the killer is not the same thing as showing that Bob is. There is also no reason to believe that this issue has to be a two horse race, or that the exclusion of one option automatically makes the other one correct. Maybe the real killer is Charlie?

This logically fallacy is also sometimes referred to as the fallacy of the excluded middle. This is worth mentioning mainly in order to avoid confusion with the Law of Excluded Middle that we looked at earlier in the book. The fallacy applies when you are only offered two options when more actually exist, whereas the law of excluded middle applies when there are legitimately only two options, for example true or false, and it is impossible for the middle or third option to exist.

Argument to moderation

This fallacy it pretty much the opposite of a false dichotomy. Rather than focusing on the two extremes to the exclusion of other options this fallacy assumes that the answer must lie in between the two options or in a combination of the two. While there are times when this may indeed be the case, sometimes the truth really does lie at one of the extremes. For example if person A believes person X is pregnant and person B believes person X is not pregnant the truth cannot be somewhere in-between these two choices, it is either one or the other. Perhaps a more realistic and nuanced example might involve different political parties and their approach to taxes. Party A says that dropping taxes by 30% is the way to solve a problem while Party B suggests raising them by 10% is the best thing to do. To conclude that the best solution therefore would be to drop taxes by 10%, thus putting the new tax rate midway between the levels proposed by both parties, is a fallacy and likely to result in neither party being able to accomplish their goals.

Proof by Verbosity

This is a favourite of conspiracy theorists of all stripes and involves throwing out so many arguments that your opponent is simply unable to respond to them all. They can then claim victory even if their opponent was able to destroy all of the arguments that they were able to get to in the time available. This works because it generally takes far longer to explain why a claim is incorrect than it does to simply state the claim in the first place. It also gives the impression that the person using this form of argument is very well versed in the topic at hand when in fact they could simply be repeating talking points without any real understanding to back them up.

> "For those who believe that JFK was assassinated by Lee Harvey Oswald I have these questions. Why did people hear shots from the Grassy Knoll? Why did Kennedy's head go "back and to the left" if he was shot from behind? How was Oswald able to pull off three shots in such a short amount of time? How do you explain the "loss" of the arrest records related to the three "tramps"? And how about the trajectory of the "Magic Bullet", do you really expect us to believe a single bullet could account for all the injuries?"

All of these points have answers, and yet to address them all takes time, and most likely more time than is available in the average debate. Even if you are able to conclusively explain why Kennedy's head would move "back and to the left", as done highly effectively on the MythBusters TV show, the conspiracy theorist can still hang onto the fact that you didn't have time to cover the "Magic Bullet" issue and claim victory.

Red Herring

The red herring fallacy involves the use of an argument that distracts from the actual point at hand. In the previous fallacy the loss of the "tramps" arrest records is a good example of a red herring. Though this does sound suspicious and makes it seem like the truth behind JFK's assassination may lie elsewhere it does nothing to address the actual facts supporting Lee Harvey Oswald

Science! It Works, Bitches!

as the sole gunman responsible. It is a distraction designed to get people thinking about something other than the evidence at hand. If 9/11 was really caused by Islamic terrorists then why did President George W Bush continue reading to school children after being told that the second tower of the World Trade Center had been hit? I don't know why he did this, I am sure he had his reasons, but this line of reasoning does nothing to argue against the evidence that Islamic terrorists were responsible for the attack, it is a red herring designed to cast doubt as to who was actually behind the atrocity. And on the subject of casting doubt I feel I would be remiss if I did not mention the great example of a red herring argument created by Trey Parker and Matt Stone for the "Chef Aid" episode of the TV series *South Park* in which attorney Johnnie Cochran unleash his infamous "Chewbacca defence". Here Cochran's character deliberately attempts to distract the jury from the facts of the case[*] with this interesting line of reasoning.

> Cochran: Ladies and gentlemen of this supposed jury, Chef's attorney would certainly want you to believe that his client wrote "Stinky Britches" ten years ago. And they make a good case. Hell, I almost felt pity myself! But, ladies and gentlemen of this supposed jury, I have one final thing I want you to consider. Ladies and gentlemen, this is Chewbacca. Chewbacca is a Wookiee from the planet Kashyyyk. But Chewbacca lives on the planet Endor. Now think about it; that does not make sense!

---

[*] In case you have not seen the episode the character Chef has an old recording that shows that he wrote the hit song "Stinky Britches" and approaches a "major record company" requesting that he be credited as the composer. The record company responds by suing Chef for harassment and hiring Jonnie Cochran to defend them.

Building your Skeptical Toolkit

Gerald Broflovski: Dammit!

Chef: What?

Gerald: He's using the Chewbacca Defence!

Cochran: Why would a Wookiee, an eight-foot tall Wookiee, want to live on Endor, with a bunch of two-foot tall Ewoks? That does not make sense! But more important, you have to ask yourself: What does this have to do with this case? Nothing. Ladies and gentlemen, it has nothing to do with this case! It does not make sense! Look at me. I'm a lawyer defending a major record company, and I'm talkin' about Chewbacca! Does that make sense? Ladies and gentlemen, I am not making any sense! None of this makes sense! And so you have to remember, when you're in that jury room deliberatin' and conjugatin' the Emancipation Proclamation, does it make sense? No! Ladies and gentlemen of this supposed jury, it does not make sense! If Chewbacca lives on Endor, you must acquit! The defence rests.

By obscuring the argument with irrelevant information the strength of the case's real facts are effectively weakened.

Tautology or Begging the Question

A Tautology, or circular reasoning or begging the question[*], is where the conclusion of the argument is also the premise of the argument. There are two slightly different ways in which

---

[*] A pet peeve of mine is when someone uses the term "begs the question" to mean "raises the question". That is not what begging the question means people.

67

something can be a tautology. The first is a statement that basically says A=B therefore A=B, for example someone might say that acupuncture works because it affects the flow of Qi energy. As acupuncture is claimed to be a way to manipulate Qi energy what is actually being said here is "manipulating Qi energy works because it manipulates Qi energy". The other form of tautology is probably a bit easier to both spot and understand. Here someone makes an argument in which their early statements are evidence for their later statements and their later statements are the evidence for their earlier ones. It might go something like this:

> Steve: How do you know the Great Leader is telling the truth?
> Barry: Because he speaks for the Ancient Spirits.
> Steve: How do you know he speaks for the Ancient Spirits?
> Barry: Because the Great Leader tells us he does.
> Steve: But how do you know the Great Leader is telling the truth?

And thus the discussion repeats itself. Interestingly there are legitimate uses for this line of reasoning as well as a number of well accepted scientific theories that are basically tautologies. For example gravity. Why do things fall down? Because of gravity. How do we know there is gravity? Because things fall down.[*] That's a tautology and yet it is a legitimate one. Let me see if I can lay that out in a slightly clearer fashion. If it can be shown that A proves B and that B proves C and that C proves D and that, finally, D proves A then we can legitimately state that A proves D and by extension B proves A. Because of this, you should always carefully examine the individual steps involved in any argument that you believe to be a tautology. If the individual steps are valid then it is

---

[*] Yes ok I know there is more to it than that, but the basic point stands.

possible that the argument as a whole may be valid even though it appears circular in nature.

No True Scotsman

The No True Scotsman fallacy is a different form of circular argument. An argument is made and when a counter argument is offered it is rejected as not being a "true" example of what is being discussed. The original example for this is still one of the best.

> Angus: All Scotsmen are brave.
> Donald: But what about Dougal? He is a coward.
> Angus: Ah but Dougal isn't a true Scotsman

Here Angus is basically making being brave part of the definition of being a Scotsman. As Dougal is not brave then he is not a true Scotsman. The definition of "Scotsman" that Angus is using is circular in that the conclusion, that Scotsmen are brave, is included in the premise, that you can only be a Scotsman if you are brave.

Argument from Final Consequences

The Argument from Final Consequences is basically an illogical line of reasoning that states that if the end result of a premise is undesirable then the premise itself must be false. Alternatively if the end result is positive then the premise must be true. One example of this that I have come across numerous times myself is the Hitler/Evolution argument.

Science! It Works, Bitches!

> "Hitler was simply following the rules of evolution, survival of the fittest, when he killed all those Jews. Evolution? More like Evilution."

The argument here is that the theory of evolution is wrong because Hitler used it to develop his master race and as an excuse to kill all the people he found undesirable, such as Jews, gypsies and homosexuals. Now even if this was true, and the evidence seems to indicate that it is not[*], it would not be evidence that the theory of evolution was incorrect. The atomic bomb dropped on Hiroshima was a result of Einstein's work and caused approximately 70,000 immediate deaths and an estimated 200,000 additional deaths due to radiation exposure. However this horrific body count doesn't mean that the equation $E=MC^2$ is incorrect.

Nirvana Fallacy

This fallacy follows nicely from the Argument from Final Consequences in that it takes the actual end result of an argument or action and compares it to an ideal and often unrealistic alternative. For example, someone might argue against giving money to a homeless person as it does nothing to address the nation's homeless problem as a whole. Alternatively, a politician's plan to help 10,000 unemployed people find work might be dismissed as it is unable to find all unemployed people jobs. Small, incremental attempts at improvement are mocked as they fall far short of a perfect solution. The person arguing against

---

[*] Hitler's views appear to have been based more upon the discredited idea of eugenics and the well-known concepts of animal breeding than upon natural selection. In fact *On the Origin of Species* was one of the books banned in Nazi Germany.

your positions does not need to come up with a better solution, they just need to point out that your solution is not perfect and then claim victory.

Argument from Authority

The Argument from Authority revolves around the idea that if the person or group making the claim are considered to be authoritative then the claim they are making must be true. This is fallacious since just because someone happens to be a figure of authority it doesn't mean they can't be wrong. If the truth of your statement is based only on the fact that it comes from an authority figure rather than any actual evidence or logic then your argument is on very shaky ground. Anyone who watches or reads the news on a regular basis will probably have come across this logical fallacy many times. Whenever you come across the statement "scientists say" or "experts claim" or "doctors recommend" then you are witnessing an argument from authority. Now this doesn't automatically mean that what the news report is saying is wrong, but if this appeal to authority is all the evidence presented in support of a claim then that is a good reason to be skeptical about it.

As with some other fallacies however there are times when the argument from authority can be a legitimate starting point from which to judge the validity of a claim. If a subject, such as the idea that the global climate is changing in part due to human causes, is said to have achieved a consensus within the scientific community then this means that the vast majority of scientists agree that the theory is correct, such as in this case where something like 97% of climate scientists agree that climate change is partly caused by humans. As such, a claim of scientific consensus can be a good jumping off point in arguing for the validity of a scientific theory.

Science! It Works, Bitches!

However, it is important to remember that the truth of the claim must always eventually come down to the legitimacy of the evidence and the logical reasoning.

Argument from Popularity

This is a fairly simple one to understand. Just because an idea is popular does not mean that it is right. Homeopathy is a multi-million pound industry and is used by many thousands of people and sold in countless shops all over the country. Should we therefore conclude that it works just because it is popular? No, it's just water people.* Another example might be that nine out of ten people voted for X, therefore X is the best thing for the country; but just because X is popular does not mean that it is the best thing to do. However once again it is possible for an argument from popularity to be logically valid, in this case when the topic at hand is, well, popularity. If 90% of the people in your group vote to go to the pub and so you decide that the group should go to the pub this is a perfectly logical conclusion, it is the one supported by the most people in the group. However if you conclude that therefore going to the pub is the best possible thing you can do you are once more committing a fallacy.

Non Sequitur

This term is Latin for "Doesn't Follow" and is used to describe an argument in which the conclusion does not follow from the premises put forward for it. Here's one you may have come across.

---

* As previously mentioned Boots chemists, the UK's biggest high street chemist, are on record as saying that they know homeopathy does not work and that they sell it because it is popular.

Building your Skeptical Toolkit

> "The fires in the Twin Towers did not get hot enough to melt steel; therefore they were brought down by explosives as part of a government conspiracy to get the USA involved in a war with Iraq."

Here we have a statement, the fires in the Twin Towers were not hot enough to melt steel, and that is correct, they weren't.[*] But the conclusion that it was therefore all a government conspiracy simply does not follow from this argument, thus the logic is invalid.

Slippery Slope

A slippery slope argument states that because you accept a milder form of the argument or proposition then you must also accept an extreme form of it as well. Not only is this not the case but also in many cases accepting the milder argument cannot even be logically argued to lead to acceptance of the more extreme form. It is also a form of the non sequitur fallacy, in that the conclusion doesn't really follow from the premise. Slippery slope arguments often pop up in highly charged political debates.

> "If you support a woman's right to choose then clearly you must also be in favour of murdering disabled children."

> "By voting in favour of health care reforms you are lending your support to the government to form death panels to decide who gets to live and who should die."

---

[*] And of course they didn't need to be, they only needed to be hot enough to weaken the steel to the point that they could no longer support the weight above them, and they were more than hot enough to do that.

Science! It Works, Bitches!

Quote Mining

Ok technically this is not a logical fallacy; however, it is something that as a card carrying skeptic you are more than likely to come across. Quote mining is the practice of taking a quote from someone who does not support your position and putting it forward as an argument in favour of said position. Personally, I think this is one of the most dishonest forms of argument that you will come across, though I will accept that it is equally likely to be done as a result of ignorance as it is of malicious intent, as many people just copy and paste these things from websites. As such I always prefer to educate rather than reprimand when I encounter something like this. One of the best and in my experience most commonly encountered examples of this comes from Charles Darwin's *On the Origin of Species*. The often quoted section states:

> "To suppose that the eye with all its inimitable contrivances for adjusting the focus to different distances, for admitting different amounts of light, and for the correction of spherical and chromatic aberration, could have been formed by natural selection, seems, I freely confess, absurd in the highest degree."

This is used to support the idea that even Darwin himself did not fully accept the theory of evolution by natural selection as valid and in fact believed it to be absurd. This quote is indeed accurate, but what makes it an example of quote mining is that it is used to support an argument that the author is not making and does not agree with. In this case this is very easy to see as in the very next sentence this quote continues:

Building your Skeptical Toolkit

"Yet reason tells me, that if numerous gradations from a perfect and complex eye to one very imperfect and simple, each grade being useful to its possessor, can be shown to exist; if further, the eye does vary ever so slightly, and the variations be inherited, which is certainly the case; and if any variation or modification in the organ be ever useful to an animal under changing conditions of life, then the difficulty of believing that a perfect and complex eye could be formed by natural selection, though insuperable by our imagination, can hardly be considered real."

It is very easy to find things people have said that seem to support almost any position imaginable when taken out of context. As such whenever you come across a quote it always pays, where ever possible, to look it up and read it in its original setting to see what the author actually meant by their statement. One thing to keep an eye out for that might help you identify when someone is quote mining are references. If someone is quote mining then more often than not they will not reference the source of the quote as this would make it too easy for someone to check out what was actually said for themselves, and this is the last thing they want. If you see a quote that states who allegedly said it but has no mention of where it is from, well, then it pays to be skeptical.

Being able to spot logical fallacies is a good skill to have as a skeptic. It can help you identify flaws in arguments about topics with which you are not that familiar and it can also help you to correct mistakes in your own reasoning and assist you in seeing where you may be accepting things for bad reasons. There are many more logical fallacies out there and during your new skeptical life you will encounter far more of them than you could

Science! It Works, Bitches!

ever wish for, you have been warned. So as we did with syllogistic reasoning let's put these new skills to the test shall we? I'll end this chapter with a few statements and all you need to do is identify the logical fallacy in each one. Good luck.

> Barry: I don't see why my tax payer money should go towards funding religious based schools.
> Steve: So you don't want to see children get a good education then?

> "If we make gay marriage legal then it won't be long until people want to marry children and animals."

> "Well I've never seen a boat or fish produce a wake like that in the water, so it must have been something else."

> "We need to have the death penalty in order to discourage violent crime."

> "I don't know why you bother having a diet coke with your Big Mac; it is hardly going to make you a super model."

The Good, The Bad and The Extraordinary
What makes something good evidence?*

Evidence is something that should be important to any good skeptic. At the end of the day evidence, of one kind or another, should form the foundation of every belief a skeptic holds. I've talked a lot about evidence in the last few chapters and so I thought that we should take a moment to look at some of the different types of evidence, examine what constitutes good evidence, what should be treated as bad evidence and if, to paraphrase Carl Sagan, whether extraordinary claims really do require extraordinary evidence.

So first off what is evidence? Well, generally speaking evidence is anything that can be used to demonstrate or validate the truth of a claim. When you gather evidence what you are doing is collecting things that are accepted as true, or have themselves been demonstrated with evidence, in order to support the truth, or validity, of your claim. Evidence can be thought of as the currency used to show that your claim meets the burden of proof or, in other words, that there is enough support for your claim to be seen as valid over the oppositional opinion, which generally tends to be that your claim is false. It is also important to remember that the burden of proof always rests with the person

---

* Here are fallacies you were hopefully able to identify at the end of the previous chapter. 1. The straw man argument. 2. Slippery slope. 3. Argument from personal incredulity. 4. Tautology (the assumption is that the death penalty helps reduce violent crime and so it is needed in order to reduce violent crime). 5. Nirvana fallacy (having a diet coke can be a good thing even if it doesn't make you a super model).

77

making the positive claim. In science for example this means that the burden of proof rests with the person presenting the claim for peer review. It is not the job of the peer review panel to disprove the claim presented to them, rather the person presenting the scientific paper for review has to demonstrate that their hypothesis is valid against all challenges.

One important distinction to make with regards to evidence is whether the piece of evidence in question is counted as circumstantial evidence or direct evidence. Circumstantial evidence is something that suggests truth, such as a fingerprint found at a crime scene. The fingerprint may be related to the crime or it may not be, as such it can only be used to infer the truth rather than directly point to it. Direct evidence on the other hand supports the truth of a claim directly. For example forensic testing could be used to conclusively show that a specific gun fired the bullet that killed our victim. This kind of evidence does not require any sort of inference on our part and can generally be treated as a fact.

As I am sure most have you have already concluded evidence can be both circumstantial and direct at the same time. For example, our fingerprint would generally be considered direct evidence that a certain person was at the crime scene at some point in time. However as we already mentioned it would only be circumstantial evidence with regards to whether that person was involved in the crime or not. It is also worth noting that numerous pieces of circumstantial evidence can be used to infer a conclusion but they never actually add up to direct evidence of something.

All that said whether direct or circumstantial some forms of evidence are simply better than others. The better the evidence you have is then the better supported your claim is. Scientists often make use of something called a hierarchy of evidence in order to evaluate a piece of evidence and see how reliable it should be considered. Now as with the scientific method the exact details of this hierarchy can vary from field to field, but generally there are a lot of similarities and as such a hierarchy, an example of which is shown below which applies most directly to the biomedical sciences, can be used to help you evaluate the reliability of any evidence presented to you, be it scientific, pseudoscientific or paranormal.

Systematic reviews and meta-analyses

In science a systematic review is generally considered to provide the best evidence you can have in support of a claim. It is a little complicated to explain so bear with me on this one. The record of all existing knowledge on a subject is known as the literature. A literature review is a study of all the critical points of that knowledge, the goal of which is to bring the reader up to date with what is currently known about the subject, generally in order to give them a starting point from which to launch further research. A literature review can be fairly broad in scope and is also prone to bias in a number of ways. For example the author may implement a form of selection bias by choosing to only include studies which support their particular point of view or with which they are personally familiar.

A systematic review is a much more rigorous process and attempts to reduce the changes of bias. It does this by focusing on a single, well-defined question and by using explicitly defined methods and a transparent approach to identify, select and critically appraise the relevant research as well as collect and analyse data from the studies that are to be included in the review. On top of that the whole systematic review process is itself reviewed to make sure that the methods used are adhered

Science! It Works, Bitches!

to and are the best possible methods for gathering the most accurate, high quality, unbiased evidence on a subject. As such, a systematic review is generally considered to represent the best evidence currently available with regard to a specific question.

As part of the systematic review process it is not uncommon for a researcher to use a statistical tool known as a meta-analysis[*]. A meta-analysis is a way of combining numerous small studies that share similar methodology and which address the same question in order to produce more statistically valid results. For example, you may have three small studies that individually seem to point in a certain direction. However when they are combined and the results examined as a whole it may become clear that the studies lack statistical significance and that the effect in question is no more than what would be expected by chance.

I must add that I am always wary when I hear about results based upon a meta-analysis of prior studies. The old saying "lies, damned lies and statistics" always springs to mind. Now while a meta-analysis can indeed be a very effective tool, and is a great way to get the results of a single very large study when you only have a number of small ones to work with, it is possible, if care is not taken or you have ulterior motives, to use them to get results that reflect your preferred outcome rather than those that best represent the truth of the matter. If the individual studies included in the meta-analysis are not of high quality and were not checked to make sure that they were conducted correctly and used a similar methodology to the other studies included in the analysis then a single bad study, or even a good study that strongly supports the desired outcome but which used a very

---

[*] Some people prefer to use the term *combined analysis* as they feel that it is clearer and easier to understand and I do agree with them on this, however meta-analysis is the more commonly used term so I will stick with it here.

different methodology, can completely throw off the results of the meta-analysis rendering it useless. As such, a good skeptic should always reserve judgement upon the results of a meta-analysis until it can be confirmed whether it was properly conducted.

Randomised Clinical Trials and Experimental Designs

I'm going to spend very little time on these two items in the hierarchy of evidence because we have already looked briefly at one of them earlier in the book and I am going to dedicate a whole chapter to the other later on. As the word clinical implies randomised clinical trials tend to be used predominately in the field of medicine to test the efficacy of some treatment or another. Experiment designs on the other hand are used widely in all fields of science and generally constitute the testing part of the scientific method. What these two things have in common is that the experiments, or tests, are designed to answer a specific question. As we know, a good scientific hypothesis is presented in the form of an "if-then" statement. A good experiment therefore should aim to present either conformation or refutation of this hypothesis. Really good experiments will also seek to demonstrate that the opposite of your hypothesis is also true. To put it another way your experiment should aim to show that:

If X happens, then Y will occur
And
If X does not happen, then Y will not occur

By testing your hypothesis in this way you can demonstrate far more effectively that your chosen cause is really having the effect you are looking for. The important thing to note with this form of evidence is that the experiment was designed, controlled and carried out to find an answer to a specific question. The literature used in a systematic review is, to a large part, made up of the results of well-constructed experiments and trials.

Cohort and Case-Control Studies

These two different kinds of studies, both generally used in medicine but with applications in numerous other scientific areas, are similar to the randomised clinical trial but with one major difference. Rather than having the person running the trial giving those involved one of a number of different treatments, or a placebo, and then monitoring the results, in both cohort and case-control studies people are selected based upon their prior exposure to the particular agent being studied or because they have a condition that the people running the study want to find out more about. Participants are then divided into at least two groups, those who have had exposure or have the condition being studied and those who don't.
In a cohort study people are selected based upon their exposure to an agent, for example exposure to large amounts of mercury. One group will have, through no action of those running the study, been exposed to the agent in question while the other group will not have been. Both groups are then followed to see if certain conditions develop more often in one group than the other. The results of such studies enable researches to say things like "exposure to high levels of mercury makes people 45% more likely to get condition X than people not exposed."
A case-control study is pretty much the same thing but in reverse. In this case the two groups are made up of people who have a particular condition and those who don't. The two groups are

then compared to see what differences exist between the two groups. For example, and sticking with the theme, it could be discovered that all the people in the group with the condition were at one time exposed to high levels of mercury while all the people in the group without the condition were not. Again this would allow researchers to draw conclusions about the relationship between certain agents and certain conditions.

Cross Sectional Survey and Case Reports

These two forms of evidence are fairly straightforward. Most of us are probably familiar with cross sectional surveys, or at least something similar to them. These are basically a series of questions, often done in interview form, answered by a sample of the population at a given point in time. Now while these tend to be better controlled and organised than the kind of surveys people stop you in the street to take, the basic principle is exactly the same.

Case reports on the other hand generally contain more detailed and extensive information than that found in a cross sectional survey, but all of it is focused on a single patient or subject. They contain, in general, a complete history of the patient's condition, how it has progressed and what treatment they received for it. From this information the researchers will be able to see that patient A received treatment X which resulted in outcome Z. Sometimes a number of case reports are gathered together into a series so that researchers can compare the treatments and outcomes of a number of different patients. If a pattern is discovered this can form the bases of a full randomised clinical trial to see if the pattern stands up to testing.

Science! It Works, Bitches!

Expert Opinion

This type of evidence is completely self-explanatory. A person, or group of people, who are considered experts in the specific field you are looking at make a statement one way or other as to the validity of a claim made about that field. As these people usually have access to the various superior forms of evidence mentioned above and are well versed in the subject their opinion is more likely to reflect what the evidence actually shows and therefore the likely truth of the matter. Rather than just relying on the opinion of one individual scientist, what is generally looked for is a scientific consensus of all the scientists working in that particular field. For example, it is often stated that something like 99.9% of scientists working in the field of biology support the idea that the theory of evolution by natural selection is the best explanation for the diversity of life on this planet. While there may be individual scientists, often very prominent and outspoken ones, who disagree with this conclusion the scientific consensus is against them and in favour of evolutionary theory.

Now it is important to remember that an expert opinion, even if it is one shared by the vast majority of scientists working in the field, should only ever be a starting point when it comes to evidence. You should always go that next step and look at the actual evidence that brought your aforementioned expert to his or her conclusion. Making an argument based solely upon expert opinion, no matter how accurate it may be, is an argument from authority, which is a logical fallacy and thus weakens your argument. Good evidence is the most important thing, not the person presenting it.[*]

---

[*] For information on how to accurately identify an expert as an expert head on over to the extras section of my website for a bonus chapter on this topic.www.andrewdart.co.uk

Anecdotal Evidence

Which brings us on nicely to the last and weakest form of evidence I want to discuss and the real reason behind this whole chapter. Anecdotal evidence is basically hearsay that results from the personal experience of the person relaying it to you or, more often, of someone they know or have heard of. Anecdotal arguments generally tend to start with something like "I did this and..."; "I know a guy who..."; "I heard of a case where..." etc. It is of course possible that the anecdote being relayed is entirely accurate, for example it could be completely true that your Uncle Joe smoked sixty a day until the day he died in an accident at 98 and never suffered any ill effects, but that doesn't mean that the evidence showing that smoking leads to lung cancer should be over turned. Anecdotal evidence is generally considered to be the weakest form of evidence and inherently unreliable and yet for many people it is this sort of evidence that will sway them to one side of an issue or the other. We see this sort of thing all the time. Someone's story about a bad experience they had with, let's say, a model of car is often enough to put people off buying one even if all the other evidence they have points to it being the best car they can afford. Likewise, a personal anecdote can affect us more than any number of facts and figures when it comes to donating to charity. Hearing that three million people are dying in a famine has less effect on us than hearing that little Kamau has to walk ten miles to get a cup of water. Anecdotes are effective and that is why charities and advertisers use them. Someone, especially a celebrity, telling you that product X made their hair shiny and dandruff free is much more effective at getting us to buy that product than a scientific report laying out in detail the evidence and experimental findings behind the claim.

Science! It Works, Bitches!

The problem is that the purveyors of the paranormal and pseudoscience know this as well. Your doctor may tell you that homeopathy has consistently been shown to have no effect in trial after trail and present you with the facts and figures to back this up, and yet if your homeopath tells you that homeopathy completely cured Mary-Sue of her cancer then it is highly unlikely that all the doctor's scientific evidence will dissuade you from giving it a try.

What often happens is that multiple anecdotes are presented to you and it is argued that this constitutes definitive evidence for a claim. Take as an example those who promote the idea that aliens are coming to Earth and abducting people in order to perform strange and invasive experiments upon them before returning them to their beds apparently unnoticed by anyone else. Almost every single piece of evidence presented in support of this claim is anecdotal in nature. UFO proponents will inundate you with story after story of strange unexplained lights in the sky, of people experiencing lost time and waking to find strange marks on their bodies, of recovered memories of abductions all of which are eerily and convincingly similar in their details. When all this evidence is presented together is can seem very convincing.

But, as many a skeptic has said before me, the important thing to remember is that the plural of anecdote is anecdotes and not evidence. While all of these stories may sound convincing when lumped together that does not mean that they should be treated as anything more than anecdotal evidence. Each claim should be looked at individually to see if it stands up on its own. When combined the anecdotal stories about some guy seeing some strange light in the sky that he couldn't identify hovering over a city and of a woman in that city who woke up to find strange marks on her body may make a compelling case for alien abduction, and yet when looked at individually they are evidence

Building your Skeptical Toolkit

for nothing more than unidentified lights and strange marks, if that.

When presented with anecdotal evidence, especially if it is coming to you by way of a third party, then your immediate reaction should be to ask for better evidence, firstly to show that the event in question happened at all and secondly to show that it happened in the way described. Only when presented with this evidence can you truly start to evaluate the subject of the claim itself.

That said anecdotes are not completely useless to a skeptic. If a number of people come forward with similar anecdotal stories it may indicate that there is something worth investigating further. A dozen people telling you that they saw strange lights in the sky is probably enough evidence to conclude that something happened and form a good starting point from which to find the truth. And that's really how anecdotal evidence should be treated, as a starting point and never as a conclusion.

Oh and as an aside if someone tries to convince you that the strange light they couldn't identify is evidence of alien visitation then you should just point out that they seem to have forgotten what the U in UFO stands for. If something is unidentified then you can't start making statements about its nature or intentions. It may surprise you to know that the two objects most commonly misidentified as alien spaceships are the planet Venus and the moon, and I'm pretty sure neither of those are trying to abduct anyone.

Extraordinary claims require extraordinary evidence

Now that we have looked at the good and the bad evidence it is time we turned our attention to extraordinary evidence, or at least the idea that extraordinary claims require extraordinary evidence. This argument, as I previously mentioned, was put forward by Carl Sagan as a heuristic or a "rule of thumb." For those who may not know, a heuristic is an easy to use tool that works in most situations but which is not intended to be strictly accurate or reliable for every situation. Some people have made the argument that extraordinary claims in fact do not require extraordinary evidence, they simply require good evidence the same as any other claim. I actually agree with this sentiment but feel it misses the point that Sagan was trying to make. It is possible that Sagan's poetic use of language may have inadvertently caused some confusion here and that a blunter, more direct, but maybe less memorable, restating of his famous phrase may clear this up. For me what Carl Sagan is saying here is 'if you want me to believe your extraordinary claim then you better have the evidence to back it up.'

Maybe I am getting ahead of myself. We should probably start by looking at exactly what would constitute an extraordinary claim. Well technically speaking there isn't really a definition of what constitutes an "extraordinary claim". That said any claim that goes against what is generally accepted as probable, has no evidence in support of it or large amounts of evidence against it or that would require the overturning of established scientific concepts and ideas in order to be considered true can be considered an "extraordinary claim". For example if I were to make the claim that I have a pet dog at home it is likely you would not find this extraordinary. However if instead I were to tell you I have a pet dragon named Barry whose fiery breath could cure cancer and who farted rainbows I expect you would consider that to be an extraordinary claim.

A pet dog is considered an ordinary claim as it is a completely probable claim that does not require us to revaluate what we know about the World. We know that there are such things as

Building your Skeptical Toolkit

dogs and we know that people keep them as pets. Accepting the claim that I am one of those people who keep dogs as pets does not require you to make a giant leap beyond what you know to be possible, nor does it require you to abandon a belief that you previously held to be true. A pet dragon however, especially one with the attributes I listed, goes against everything we know about the World and what is probable. You have no reason to accept that dragons are even real let alone that I keep one as a pet and doing so would probably require you to abandon certain other things that you previously accepted as true. The fact that in this instance neither claim is true is neither here nor there, as the accuracy of the claims does in no way effect reality with regards to the existence of dogs and dragons.

Other extraordinary claims require us to do away with what we know about science before they can be accepted as true. This does not automatically mean that the claim in question is false, only that if we are to dismiss an idea that up until this point has been supported by all of the available evidence in support of this new claim then the evidence showing that this new claim is correct and that the old claim is wrong needs to be of a very high standard. Neurologist and influential skeptics Dr Steven Novella put it like this:

> "We do not and cannot know everything. We will never obtain 100% metaphysical certitude regarding any scientific question. But there are some aspects of nature that have been verified to such an outrageously high degree that, barring an equally outrageously high degree of evidence to the contrary, treating them as anything other than a rock solid fact is absurd bordering on insanity."

Science! It Works, Bitches!

There are some areas of science that are so well supported by evidence that to state that they are untrue would itself be an extraordinary claim. When we looked at the scientific method we covered what goes into verifying that a scientific claim is true. You have to run experiments that support your hypothesis. You have to put your conclusions forward for peer review so that other people can recreate your experiments to see if they get the same results. Only once your results have been replicated many times and shown to be accurate is a hypothesis granted the provisional acceptance of being true. Accurately following the scientific method and passing the peer review could be classed as extraordinary evidence for a claim and any counter claim would not only have to present equal evidence in support of itself but also enough evidence to show definitively why the initial claim is wrong.

While it is true that looked at on its own all evidence is ordinary evidence it is the sheer amount of evidence required to show that something is true that can be classed as extraordinary. Another way of thinking about it is to realise that for most every day claims we already have what would constitute extraordinary evidence were we starting from scratch. If we think about it this way it is clear that the evidence to prove I have a pet dog would be exactly the same as to prove I have a pet dragon. I would simply have to show you the pet in question. However, one claim is considered ordinary and the other extraordinary because the idea of dogs as pets already has a preponderance of evidence in its favour while dragons as pets does not. A living breathing dog that we could examine, test, measure etc may not seem like extraordinary evidence but that is because we have already long accepted the idea that dogs are real. A living breathing dragon on the other hand would strike most of us as extraordinary even though, obvious differences aside, a dog and dragon would just be two different kinds of animal.

At the end of the day whether you accept the idea that extraordinary claims require extraordinary evidence or not, or even that a specific claim is itself extraordinary, what it all comes down to is the evidence. The evidence presented to accept something as true has to be good enough to support your claim and, if needed, good enough to refute the counter claims. If you want me to believe your claim, extraordinary or otherwise, then you better have the evidence to back it up.

Science! It Works, Bitches!

Skeptical Sidebar #3
The File Drawer Effect

In the previous chapter we looked briefly at the systematic review process and meta-analysis as being some of the strongest ways to look at the evidence for a given claim. I also mentioned a couple of failings with this approach and I want to take a quick look at another such problem that can have an influence on what the evidence appears to show and which you are very likely to come across as a budding skeptic.

The file drawer effect, also often referred to as publication bias, happens when research into a given subject is only submitted to be published if the findings are statistically significant. Research that shows little or no effect or that supports the current thinking on the subject can often end up not being published and instead filed away in a drawer somewhere. Just to be clear here we are not talking about research that finds a negative result, as negative results can still be statistically significant, but rather about research that simply doesn't find an effect. For example this might be research into psychic powers that shows the "psychic" performs no better than expected by chance under controlled conditions. Or it could be research on a new drug that is shown to be no more effective than a standard placebo, something we will look at in more detail later in the book. As studies like this tend to support the null-hypothesis, something else we will look at in a bit, they are far less likely to be published than research that shows a distinct effect one way or another.

Now obviously this is a problem as in order to carry out an effective systematic review or meta-analysis you need to look at all the results found on the subject, positive, negative and, equally if not more important, neutral. Let's say that fifty experiments are carried out into whether people have psychic powers or not. Two of these experiments find a positive result and are published, the remaining forty-eight however find no psychic effect and that the results match those expected by chance and so are filed away in a drawer somewhere. When someone comes to review the literature on this psychic effect they find the two research papers reporting a positive finding but not the forty-eight that show no effect and as such any conclusion they draw is highly unlikely to be representative of all the research into the subject. And when the same thinking is applied to the effectiveness of medical treatments I am sure you can see the possibility for harm here.

So what causes this? Why is it that papers that find no effect are far less likely to be published than those that do? Well one possible explanation is the sad fact that in many areas of academia there exists a "publish or perish" culture whereby academics feel pressure to publish in order to have a successful career. Certainly those seeking tenure or a senior position within their department are expected to publish papers on a regular basis and research shows that the more competitive the environment the more likely it is that the papers submitted to journals will contain only significant results. The truth is that research that shows a statistically significant result is far more likely to be published by scientific journals than research that

Building your Skeptical Toolkit

shows a neutral result, if for no other reason than it is more interesting and therefore more likely to sell copies of the journal, and as such this is the kind of research that tends to be submitted to the journals in the first place.

One possible solution to this problem, and one that a number of countries have or are in the process of implementing with regards to clinical trials, is to make it mandatory for all trials to be registered with a central database before they are carried out. This registration, which would include a clear record of the goals of the trial, would make it possible for people to get a clearer picture of all the research being carried out in a given area and reduce the file drawer effect by maintaining a record of trials whether they are published or not. Also by keeping a record of the goals of the trial it is possible to compare these initial goals with those stated in the final published paper, thus reducing the effect of another publishing related problem, that of people changing the nature of the trial after it has been conducted in order to produce more favourable or statistically significant results. A paper published in the Journal of the American Medical Association in 2009 found that even amongst those trials that registered beforehand almost one in three changed their stated goals before final publication, so this a problem that needs addressing.

As a skeptic it is always worth keeping the file drawer effect in mind, especially when faced with scientific papers that seem to support something that sets off your skeptical radar. For every paper that indicates the presences of psychic powers or the effectiveness of homeopathic treatments there could well be dozens of others that found no effect whatsoever and were simply never published.

Science! It Works, Bitches!

Skirting the razors edge
Occam's razor and Hume's maxim

At the end of the chapter on evidence we looked at the heuristic "extraordinary claims require extraordinary evidence". Well now is as good a time as any to take a very quick look at two more heuristic statements that you will probably find useful, Occam's Razor and Hume's Maxim. Now I suspect you have probably all heard of the former but you may be less aware of the latter. Both of these heuristics deal with very similar issues and are both ways in which you can critically assess the truth of a given issue when you have very little else to go on other than the statement of the person making the claim.

Occam's Razor

While most people have heard of Occam's Razor, also known as the law of parsimony, it is surprising how often it is misused or misquoted. Most people think of Occam's Razor as being "The simplest explanation is usually correct." Now while this version is essentially correct it does miss out an important point. A better way of wording it would be "The explanation that introduces the fewest new assumptions is usually correct." Let me give you an example that hopefully will make the important difference between these two versions clear.

Let's say that a series of strange events take place around the area in which you live. Crop circles appear overnight in fields. Dead cows are found with their insides apparently ripped out.

Science! It Works, Bitches!

People report missing time and strange visitations while they sleep. Now a seasoned skeptic would probably put forward a number of explanations for these events. The crop circles were made by people late at night with a piece of string and a wooden plank. The hollowed out cows are simply the result of expanding stomach gases and tiny carnivorous scavengers. The missing time could have many explanations, the simplest being that they simply lost track of time, and the strange visitations are most likely the result of waking dreams. However the UFO believers have a different take on the matter. They claim that the skeptic's have different answers for everything, while they just have one solution. Alien visitation. They then misuse Occam's Razor to claim that because their explanation is the simplest then it must be correct.

However if you use the correct version of Occam's Razor, that being "The explanation that introduces the fewest new assumptions is usually correct" then you come to a different solution. In order for the UFO proponents to be correct they have to introduce a number of new assumptions. Firstly, that intelligent alien life exists. Secondly, that it has the means to travel across the vast distances of interstellar space. Thirdly, that it has done so and has come to Earth and fourthly that the events described are the kind of things that these aliens would do once they got here. The skeptical explanations may be numerous but none of them introduces any new assumptions. We know that people can and have made crop circles. We know what happens to dead cows if they are left to nature. We understand the various causes of missing time and we understand the effects of waking dreams. Therefore, if we apply Occam's Razor correctly it becomes clear that the skeptical explanations, rather than the single explanation of the UFO proponents, are more likely to be correct.

Building your Skeptical Toolkit

Hume's Maxim

Hume's Maxim, named after the famous Scottish philosopher David Hume, is very similar to Occam's Razor. However, whilst Occam's Razor deals with the assumptions that the claim is making, to my mind, Hume's Maxim deals more directly with the claim itself. Hume described his maxim like this and gave the following example:

> "That no testimony is sufficient to establish a miracle, unless the testimony be of such a kind, that its falsehood would be more miraculous than the fact which it endeavours to establish.
> When anyone tells me that he saw a dead man restored to life, I immediately consider with myself whether it be more probable, that this person should either deceive or be deceived, or that the fact, which he relates, should really have happened. I weigh the one miracle against the other; and according to the superiority, which I discover, I pronounce my decision, and always reject the greater miracle. If the falsehood of his testimony would be more miraculous than the event which he relates; then, and not till then, can he pretend to command my belief or opinion."

Now I admit that the 18th century English can be a bit tricky to follow in places so allow me to paraphrase. Hume's Maxim basically states that testimony alone, as we covered in the previous chapter, is never enough to establish the reality of a miraculous or supernatural event, unless the idea of the testimony itself being false would be more amazing than the claim that it is making. Hume simply advises us to weigh up the odds. Is it more likely, to use his own example, that a man has come back from the dead or that the person reporting this event is either simply mistaken, has been deceived in some way themselves or is trying to purposefully deceive you? Whichever option is the least likely is the one that can be dismissed. Now it is worth mentioning that Hume's Maxim formed the basis for Carl Sagan's "extraordinary claims require extraordinary evidence" idea. Like Sagan, Hume was simply pointing out that if you are going to make a miraculous claim then you'd better have the evidence to back it up.

By using these two heuristics together you can very quickly weigh up the likelihood of a claim being accurate, even if you have no more evidence to go on than the claim itself. First, weigh up the odds. Is it more likely, knowing what you do about the World, that the claim is correct or that the person making it is mistaken in some way? Then try to come up with an explanation that does fit with what you know about the World and use Occam's Razor to decide whether your explanation or the one being offered to you introduces the most new assumptions. Do this and, in the words of The Who, you won't get fooled again. With that in mind let's get back onto the subject of science.

Is there a Doctor in the house?
A quick guide to clinical trials

While we have already looked at the scientific method in general there is one application of this kind of scientific, evidence-based thinking that deserves a closer look, and that's the Randomised Controlled Trial, better known as a Clinical Trial. As a skeptic one of the more common situations where you will encounter pseudoscientific thinking is with regards to alternative medical treatments. In the UK there is a big problem with these kinds of treatments being looked upon as just as valid as conventional, evidence based medicine. They receive funding from the government, are praised by the media and are even supported by members of our Royal family. Very few people actually understand the reality of how these alternative modalities work, or rather don't work, and the potential risks involved in seeking this sort of treatment over standard medical practices. Musician and comedian Tim Minchin summed up the situation nicely in his ten minute beat poem "Storm":

> "By definition", I begin,
> "Alternative Medicine", I continue,
> "Has either not been proved to work, or been proved not to work.
> Do you know what they call 'alternative medicine' that's been proved to work?
> Medicine."

Science! It Works, Bitches!

One of the reasons that people tend to fall for the bright lights of alternative medicine is that they lack an understanding of how medical treatments are tested to see if they are safe and effective. Most form of alternative medicine rely on anecdotal evidence to support their claims and as we saw earlier this is one of the least reliable forms of evidence, especially when it comes to your health. When it comes to real medicine something more concrete than the word of someone who had, or believes they had, a positive experience is required, and this is where the clinical trial comes in.

Clinical trials are used within medical research to address questions about health, illness, and treatment options, mainly with regards to the safety and efficacy of a specific treatment. Clinical trials can only be carried out once a treatment or product has already produced enough evidence to demonstrate the quality of the product and that it is "non-clinical" safe, which usually means it has been shown to be safe in animal trials. It also has to be shown that the trial is ethical before it is allowed to be carried out and that you have the full written consent of the people involved. Even then the mechanics of the trial can vary widely depending on what is actually being tested and at what stage in development the product is at. Taking all of these things into account what I am going to outline below is a somewhat idealised version of how a clinical trial should be conducted. Remember that clinical trials still sit within the overall scientific method that we looked at earlier and would fit, unsurprisingly, into the testing part of that method.

Ok, so let's assume that our research has reached the stage of human testing and that the trial we are planning to carry out has been signed off upon by the ethics committee of whatever country we are carrying it out in. What do we do next?

Building your Skeptical Toolkit

Well first of all we need to make sure that our trial is big enough to be statistically significance. This is known as having a large sample size. The reason for this is fairly obvious. Let's say we do a trial on three people and one of those three is completely cured by our new procedure. Now does this mean that our procedure cures one in three people? The answer is we don't know. That one person could have been a complete fluke. Let's say we repeat the trial, this time with 3,000 people. Now if the procedure really does cure one in three we should have 1,000 people completely cured by our treatment. But this time only two people are cured. Clearly having a larger sample sizes allows us to better identify unusual outcomes than trials with a small sample sizes. The best trials often involve tens of thousands of people.

So let's say we have 20,000 people volunteer for our trial. The next step is to go through them and make sure that our test sample accurately reflects the population at large and that we remove any chance of selection bias. It would be very easy to get great results from our trial if all the people in it were athletes in their twenties. These people are more likely to recover from or better survive a condition anyway and as such our trial would not accurately reflect the effect the treatment would have on the general population, which includes children and old people. By removing the risk of selection bias we reduce our sample size but end up with a trial that better reflects the effects of the condition in the real world.

So we now have a carefully selected sample size of, for want of argument and easier maths, 9,000 people. We now have to divide these people up into a number of groups that will each receive a different form of treatment. This is done randomly, again to remove the chance of any kind of selection bias. In this example we will randomly put 3,000 people into each of the following three groups. The first group will receive the new treatment, the second will receive an existing treatment for the condition we are studying and the third will receive a placebo[*]. Now I am going to go into the subject of placebos more in the next chapter, as there is a lot to be said about them, but here I am simply going to say that a placebo in a clinical trial is a faux-treatment that is generally delivered in the same way as the actual treatment but which has no actual effect on the condition in question. Actually, I will just add that placebos are not always completely inactive. Let's say that both the new treatment and the existing one are known to cause nausea. In that case the placebo would also be something that causes nausea but which otherwise has no effect on the condition being tested. The reason for this brings us onto the next step in our clinical trial.

Our sample size has now been randomly assigned to three different groups and we are almost ready to start treatment. Before we do so however we need to make sure that our test subjects are properly blinded. Now this doesn't mean we blindfold them, though the overall effect is basically the same. Blinding means that no one knows whether they are receiving the new treatment, the old one or the placebo. As with the example above this can mean using a placebo that replicates the known side effects of the actual treatment or, if the treatment is administered by injection, that the placebo is also administered in the same way. The reason for blinding is to remove, as best as possible, the psychological effect that knowing they are getting a new treatment will have on a patient. Research has shown that if patients are told that a treatment will have a certain effect then they are more likely to report that effect, even if they are in fact

---

[*] It is worth noting that there are some ethical implications with regards to giving patients a placebo in lieu of an actual treatment for their condition. As such things are often not a simple in the real world as I present them here, but the basic practice and procedure is the same.

Building your Skeptical Toolkit

just receiving a placebo. People tend to assume that new treatments are better than old ones and that both new and old treatments are better than a placebo. As such if they are not blinded then there is a chance they will report improvement of their symptoms, or not, because they believe that is the answer the doctor is expecting, rather than because their symptoms are actually reduced.

Which brings us on nicely to the reason why good clinical trials are generally double blinded. This means that not only do the patients not know which of the three treatments they are getting, the new, the old or the placebo, but neither do the doctors who are giving it to them. By blinding the doctors you remove the chance of them accidentally, or even intentionally, influencing the patient in any way with regards to the effectiveness of a treatment or letting it slip which of the three treatments they are actually getting. For example, a doctor who knows he is giving a placebo may tell an excited patient not to get their hopes up. Likewise, a doctor giving a new treatment in which they have a lot invested may encourage a patient into reporting the results they want to hear. As such in clinical trials treatments are generally encoded to remove the risk of psychological bias. For example, patients in Group 1 may receive Treatment A. However, neither the patients nor the doctors know if Treatment A is the new treatment, the old one or the placebo.

Everything is now set and we can start our trial. The important thing here is to make sure that the trial lasts long enough to give statistically accurate results. It is very tempting to stop the trial as soon as you get any positive results for the treatment you are testing but this is not the way to get accurate results. For example let's say that our new treatment has very positive effects within the first week of our trial. As this is the result we want it would be very tempting to just stop testing here and announce that our treatment works. However let's say that the next week all those people who showed early improvement suddenly break out in a terrible rash and start having problems breathing. For this reason the best trials, as well as being very large, also last for a long time, often a number of years or even a few decades.

Right, so it is now five years later and we have collected the results of our trial. What next? Well the results need to be reviewed by statisticians to see what they show and once again in really good trials, like our one, this is a blinded process. A triple blinded study means that the people reviewing the results don't know which treatment is which either. All they know is that the patients in Group 2 receiving Treatment B seem to have a better rate of recovery than the groups receiving treatments A and C. The results are also reviewed to see if they are statistically significant. Yes Group 2 had a better recovery rate but is it large enough to indicate that their treatment was more effective or is it simply within the limits you would expect to see by chance? Let's say, for arguments sake, that Group 1 had a recovery rate of 36%, Group 2 38% and Group 3 32%. Are those differences statistically significant or are they just the kind of background noise you would expect to see in any trial involving us wonderfully varied human beings?

Once the results have been evaluated and it has been confirmed if they have any statistical significance the blinding is finally removed and we find out which treatment is the most effective. In our case it was Treatment B, which turns out to be the old treatment. Treatment A, our new treatment, was more effective than Treatment C, the placebo, but less effective than the form of treatment currently in use. As such we will carry on using Treatment B and maybe go back to the drawing board with Treatment A to see if we can make it more effective. Actually in the case of our trial both the current treatment and the new treatment were both very close to the results recorded for the placebo. As such it could be concluded that neither treatment actually has any kind of statistically significant effect over that of the placebo effect.

Of course, as I said earlier, the clinical trial represents just one way of carrying out the testing stage of the scientific method. We now have to continue with the next stage, that being reporting our findings and allowing others to evaluate our work for errors and, if inclined, develop their own trials in order to test our results. So as we can see medicine, as with all other areas of science, improves by constantly evaluating the effectiveness of treatments in the face of new evidence and dismissing those treatments that are less effective, or which simply do not work, in favour of those that do. Once again the power of science is found in its willingness to let go of old beliefs if they conflict with the evidence and in constantly trying to improve itself rather than clinging dogmatically to old ways of doing things.

Science! It Works, Bitches!

Do you take the red pill or the blue pill?
The placebo effect

As doctor and author Ben Goldacre said in his book Bad Science, the placebo effect is one of those subjects that seems simple at first but about which you could write entire books and not even scratch the surface. But don't worry; I'm going to try to keep this as short as possible and just cover the basics as well as a few interesting little facts.

As I mentioned previously a placebo is a sham medical intervention or treatment that is designed to convince the patient that they are receiving an actual treatment for a condition or illness, but which has no actual effect upon said condition at all. Placebos are generally inactive substances such as "sugar pills" and saline solution. However there are times when, in order to be effective as a placebo, it is necessary for the placebo to replicate the known side effects of the treatment being tested so that the patient, as well as their doctor, remains unaware of whether they are receiving an actual treatment or not. It is rare these days for placebos to be given as a form of treatment outside of clinical trials due to the Hippocratic Oath and the importance of honesty in the doctor-patient relationship. It is worth noting that such considerations do not seem to apply within the realm of alternative medicine, which, as we shall see, relies heavily upon the placebo effect in their treatments and thus requires practitioners to convince patients they are getting something when in fact they may only be getting, and in some case we are talking literally, a sugar pill.

Science! It Works, Bitches!

Which brings us to the question of why, if they have no effect, are placebos ever used? Well the answer is that placebos are used in clinical trials so that the people performing the trial can rule out the placebo effect. The placebo effect is a measurable, observable, or even felt improvement in health that is not related to the treatment being given. To put it another way the belief that you are being treated is often enough to make you feel like your condition is improving, even if you are in fact not being treated. This is why placebos are used in clinical trials. Without them it would be very difficult to tell if the feeling of improvement a patient has after receiving a new drug is caused by the drug itself or simply by the patient's belief that the new drug is going to make them better. It is important to note that, except in very rare instances, placebos do not actually have any effect upon the actual condition or illness but rather affect subjective symptoms such as pain and fatigue. As such a placebo may make the symptoms of something like cancer feel less extreme but will have no effect upon the cancer itself, while on the other hand they may indeed work wonders with regards to treating something like back pain that can be based upon the patient's subjective experience rather than a physical symptom.

There is a flip side to the placebo effect known as the nocebo effect. This is where the patient experiences a negative effect from a placebo treatment because they believe that the treatment will have a negative effect[*]. In both cases the patient's view of the treatment is what causes the results, not the treatment itself. If they believe the treatment will have a positive effect it is more likely that the placebo will indeed result in some sense of improvement. On the other hand if they believe the treatment will make them sicker or that it has serious side effects

---

[*] This is separate from the use of placebos that actually cause negative symptoms that replicate the side effects of the actual treatments.

Building your Skeptical Toolkit

then it is possible the patient will experience those side effects even without getting the actual treatment. The effects of placebos are very real and can indeed affect the overall outcome of any treatment. This is why it has to be considered when conducting clinical trials and also why it is generally considered unethical to use placebos in everyday treatment of patients, as the effectiveness of the treatment is not found within the treatment itself but in the patient's mistaken understanding of its effectiveness.

There have been a number of trials conducted upon placebos themselves that have produced some very interesting results. For example, it has been found that the invasiveness of the treatment has an effect on the perceived results, even if the treatment itself is completely ineffective. Two pills have more effect than one and an injection has more effect than pills. Packaging and presentation also has an effect. Green sugar pills do a better job at treating anxiety than red pills. Sugar pills presented in fancy, expensive looking packaging have more effect than the same pills presented in plain cardboard boxes. The same applies to pills labelled with known brand names as opposed to generic brands. This actually applies to real medicines as well. Brand name aspirin and ibuprofen are reported as being more effective and quicker acting than unbranded medication with the exact same active ingredients. People report that they work better because they believe they work better, which in turn actually causes them to work better. This is an instance where a tautology can actually work for you. If you believe that it is going to work then there is a very good chance that it will, at least to some degree. Personally, I know that when I take a pill for a headache I generally start to feel better immediately even though the medicine hasn't had enough to time take effect.

Science! It Works, Bitches!

Which brings me on nicely to alternative medicine. Time and again treatments such as homeopathy and acupuncture have been shown to have no more effect than that caused by a well-presented placebo. When you go for alternative treatments it is very likely that the practitioner will be able to spend a lot more time with you than your regular GP. They will also generally see you in a much more relaxed environment, soft music playing, incense burning in the corner, and will offer you comforting words, assuring you that whatever problem you have can easily be treated by them and that you have no need to worry. It is not at all surprising that in this environment the placebo effect can indeed be very pronounced. In fact many doctors have petitioned to incorporate many of these elements into modern medicine, in effect combining the old fashion caring family doctor feel with the best modern medicine has to offer. Unfortunately, time and resources often make this much more difficult for actual doctors than it is for those who rely almost completely upon the placebo effect for repeat business. Allow me to give you an example.

In 2009 a number of main stream media outlets ran a story that announced how sham "toothpick" acupuncture is just as effective as actual acupuncture. This story was based upon the results of a study, published in the Archives of Internal Medicine, in which acupuncture was tested to see if it is effective in treating lower back pain. The test subjects were divided into three groups. The first received traditional acupuncture, which involves placing needles at specific acupuncture points on the body that are supposed to corresponds to the pathways through which Qi (life energy) flows. The second group also received acupuncture, only this time the needles were placed at random so that they avoid any recognised acupuncture points. The third group, the

Building your Skeptical Toolkit

important one, was the placebo group[*]. This group received sham acupuncture. Toothpicks were used to simulate the feel of the needle on the skin, but importantly did not actually pierce the flesh. And again these were placed in such a way as to avoid any recognised acupuncture points. And the results? Well it was found that all three groups reported almost exactly the same results. The media immediately jumped upon this as evidence that not only does acupuncture work, but sham acupuncture works as well. However, as I am sure you have already worked out for yourself, this trial shows nothing of the sort. What it shows is that acupuncture, of any variety, has no more effect than that explained by the placebo effect. In other words it simply doesn't work. The two most important parts of acupuncture, the placement of the needles and the needles actually piercing the skin, can be removed from the equation and you get the exact same results. So in short, this study shows that not getting acupuncture is as effective as getting acupuncture.

The vast majority of alternative medical treatments rely heavily upon the placebo effect. It is completely true that after spending a relaxing half hour listening to pan pipes and breathing refreshing scents while your alternative treatment practitioner listens intently to your woes and happily informs you that taking just two of his magical sugar pills a day will heal you of all your ills, that you can indeed feel much better than you did before. The important thing to remember is that, while placebos can be effective against subjective symptoms, they do not treat actual underlying causes of illness. If you are feeling unwell remember that just because a trip to your local homoeopath or acupuncture

---

[*] There was actually a forth control group that received no treatment, other than whatever they were getting from their regular physicians. However, as this group is not relevant to this discussion at hand I have excluded them from this write up of the experiment.

Science! It Works, Bitches!

technician may make you feel better it doesn't mean you are in fact better. Always consult a doctor first. They may not offer easy answers but at the same time no amount of placebo treatment will remove a tumour or fix a sick heart. The danger comes when the patient believes that an alternative treatment is working because it makes the "feel" better and so they put off seeking actual treatment until it is too late.

Stop global warming, become a pirate
Correlation and Causation,
Possibility and Probability

While we are briefly on the subject of alternative medicine there are two more quick things I want to touch on due to the way in which alternative medicine, as well as pseudoscience and the paranormal as a whole, replies upon them to a lesser or greater extent. The first is the relationship between correlation and causation and the second the equally fraught relationship between possibility and probability. Let's take them in order.

Correlation and Causation

Scientists spend a lot of their time looking at the relationship between cause and effect, be it cosmologists looking for the cause of a super nova or a medical doctor trying to find the reason behind a patients symptoms. Sometimes these cause and effect events are closely related, be it spatially, temporally or both, and are therefore obvious and easy to spot. However this is not always the case and there are times when the cause of something may be greatly separated from the effect, again either spatially, temporally or both. Science takes this into account and by rigorously following the scientific method and working with the best evidence available it does its utmost to make sure that the answer it comes up with is the correct one and not just the easiest one.

Science! It Works, Bitches!

Unfortunately, in everyday life most of us don't do this and us humans seem to be very much inclined to assume that if two events are closely related in space and time then they must be causally related as well. This problem occurs because the vast majority of people have difficulty in telling the difference between correlation and causation. Correlation, as defined in the Oxford English Dictionary, is the "mutual relation between two or more things" whereas causation is described as "causing or producing an effect". Just because two events may correlate, there being a mutual relationship between the two, does not mean that one event is the cause of the other. Bringing this back to alternative medicine, just because your cold went away after you received acupuncture does not mean that the acupuncture made your cold go away.

The idea that correlation equals causation is a logical fallacy known as cum hoc ergo propter hoc, which is Latin for "with this, therefore because of this". This fallacious line of reasoning can be thought of in this way:

> A occurs in correlation with B
> Therefore A causes B

Just because event A was seen to happen in correlation with event B, be it at the same time or at the same place or both, does not mean that event A was the cause behind event B. Now while it is possible that event A did indeed cause event B the correlation equals causation argument is a logical fallacy because there are at least four other possible alternatives.

B may be the cause of A

A simple example makes this one very easy to understand.

> The more doctors treating the sick, the more sick people there are going to be in hospitals
> Therefore doctors cause sick people

It is true that there is a strong correlation between the number of doctors involved in treating the sick and the number of sick people they are treating, but of course this doesn't mean that the doctors are the cause of the increase in sick people. In fact the opposite is more likely to be true, that the increase in sick people results in the need to bring in more doctors. So in this case event B is the cause of event A.

The third factor C is the cause of both A and B

Again this is very easy to understand. In this case a third, possibly hidden, event is the cause of both event A and B. This third event is known as a common-causal variable. Here's an example that you may have experienced in your own life.

> When I eat ice cream I get acne
> Therefore eating ice cream gives me acne

Now of course we can't rule out that this conclusion is entirely correct but we shouldn't jump to it just because the two are strongly correlated. We need to make sure that we have ruled out the possibility of a third hidden cause for these two events, in this case stress or anxiety. When stressed many of us may head straight for the freezer for some ice cream, or perhaps chocolate is your comfort food of choice. However, stress is also a common cause of acne and as such while our eating of ice cream may

Science! It Works, Bitches!

correlate closely with our outbreak of acne it is just as likely that the stress we are under is the cause of both.

Coincidence

This one doesn't really need an explanation, as it is fairly obvious. However, it does give me an excuse to invoke one of the central tenets of parody religion the Church of the Flying Spaghetti Monster.

> As the number of pirates has decreased the global temperature has increased
> Therefore global warming is caused by the lack of pirates

Ramen my Pastafarian brothers and sisters. Of course no one really takes this seriously, at least I hope they don't. While it is in true that the average global temperature has increased in correlation to the decrease in the number of pirates around the World this is simply a coincidence and there is no causal relationship between the two, mores the pity, Yarrgh!

A causes B and B causes A

Unfortunately I can't think of an amusing example for this one and so I will have to use a scientific one.

> As you increase the pressure of a gas the temperature also increases
> Therefore increased pressure causes increased temperature

In the case of this example the relationship between correlation and causation is completely accurate. If you increase the pressure

Building your Skeptical Toolkit

of a gas then the temperature does indeed increase as well. However, the inverse is also true. If you increase the temperature of a gas then the pressure will also increase. This is known as the ideal gas law, which for those of you who care can be written as $PV=nRT$. This describes the fact that there is a direct relationship between temperature and pressure. Increase one and you automatically increase the other and so in this case the correlation equals causation argument works both ways.

When faced with a new claim it is always important to keep in mind that correlation does not necessarily imply causation. As I said many pseudoscientific claims rely on our very human issue with confusing correlation for causation to convince us that they work. For example let's say that you have a friend that swears blind that mega dosing with vitamin C is the best way to cure a cold. She tells you that whenever she is feeling her worst she takes large amounts of vitamin C and the cold goes away. Ignoring for the moment that this would constitute anecdotal evidence which, as we already know, is the least reliable form of evidence, it is likely that your friend is also committing the correlation equals causation fallacy. Just because she takes large amounts of vitamin C when she is feeling really bad does not mean that this is what cures her cold. In fact, it is more likely that something far simpler is happening here, an effect known as returning to the mean. Colds, along with many other things, happen in generally predictable ways. You feel ok, you get a cold and start to feel bad, you get really bad and then you start to get better. When you are at your worst then most times the only direction things can go is towards you feeling better. Take any treatment when you are feeling your worst and, no matter how ineffective it may actually be, it will appear to make you feel better as your body naturally starts to return to its normally healthy state. That aside let's move on shall we.

Possibility and Probability

The relationship between things that are possible and things that are probable is another one that most people tend to have a problem with. If we are honest we can admit that pretty much anything, no matter how crazy, is possible. Ghosts could really be communicating with psychics from "the other side". UFOs could really be abducting people from all over the place without anyone noticing. Homeopathy could really be effective for treating numerous medical conditions...actually scrap that last one as it is too crazy even to use as an example. The point is that no matter how out there a claim may be it is possible that it is true, and that goes for everything that proponents of the paranormal and pseudoscience put out there as well as that guy who tells everyone that he's Napoleon. After all, it is possible he's telling the truth.

So with that in mind what we should really care about is not what is possible, as that encompasses almost any claim a person could make, but what is probable. While it is possible that guy is really Napoleon is it really probable that he is?

Before we go on I feel that I should include a quick caveat in the whole "anything is possible" argument. While all the examples I have given so far are all indeed possible they all have one thing in common. All of them conform to the laws of logic that we talked about earlier and it is these laws, as well as the principles of mathematics, that define the limits of what is possible. For example, it is not possible to have a three-sided square. This is because the feature that defines a square is that it has four equal sides. As such a three-sided square would not be logically consistent. The same goes for mathematics. 1 + 1 cannot equal 5 because we define the next number after 1 as 2 and so therefore it would not be logically consistent and so not possible.

Building your Skeptical Toolkit

However, in the everyday world the issues we tend to run into with regards to possibility and probability tend to readily conform to the laws of logic. For example, I could state that tomorrow the Sun will rise in the west and, ignoring the fact that technically the Sun doesn't rise at all, I would still be logically consistent and as such it is possible that this could happen. Which brings us back to the probability of it happening.

When trying to assess the probability of something being accurate or true you will not be surprised to hear that we once more must start by looking at the evidence. A good way to do this is to make use of something called Bayes' Theorem, though I'll admit I am probably about to use it in a way that would make most mathematicians weep. Imagine that I am going to flip an honest, nonbiased coin three times. I flip it once, it comes down heads, and I find myself wondering what the probability is of me getting three heads in a row. This is where Bayes' Theorem comes in handy. I start with the evidence I have to hand, which in this case is the fact that it is an honest coin, and as such it has a 50% chance of landing on heads on any given flip, and that it has already landed on heads once. I can use this information to work out something called a prior probability, which in this example is the probability of flipping a coin and it landing on heads three times in a row. I do this by simply multiplying together the odds of the coin landing on heads for each of the three flips of the coin, which gives us a prior probability of three heads in a row of 12.5%. I can now use this information to calculate what statistics calls the "likelihood". Here the likelihood is the probability that I will flip heads given that I will flip three heads in a row, which I am pretty sure you can all work out to be 100%. Now that I have all this information I can use it to calculate the answer to my question, what is the likelihood of me flipping heads three times in a row given that I have already flipped heads once. In Bayesian terminology this is known as the posterior probability, which for those of you playing along at home works out to a 25% chance of flipping three heads in a row if I have already flipped heads once.

And that is pretty much the basics of Bayes' Theorem.[*] You start with a prior probability of something being true on its own, you then examine the available evidence and use this to calculate a posterior probability, the probability of it being true given the available evidence. We can apply this method to anything that is not logically necessary, such as a square having four sides, and where there are at least two probably outcomes, as with our coin flip. Which brings us back nicely to the topic of alternative medicine.

Let's go back to the use of acupuncture for the treatment of lower back pain that we covered in the previous chapter. Before we look at any of the evidence we might decide to apply a very generous prior probability of acupuncture being an effective treatment for lower back pain of 50%, that is to say that it is just as likely to be effective as it is likely to be ineffective. We can then look at the evidence, which in this case would be the study published in the Archives of Internal Medicine, and see how this affects our probability level. Unfortunately for acupuncture the evidence shows it to be no more effective at treating lower back pain than a well administered placebo, and so with this information we can come up with a posterior probability of it being an effective treatment that is far lower than our initial 50%.

Now of course when faced with a new claim we can't always sit down and calculate the Bayesian mathematics then and there, however what we can do is follow the basic principles of Bayes' theorem. Evaluate the claim on its own and then examine it by the light of the available evidence to see if the claim is more or less likely to be true given that evidence. What is the probability

---

[*] Cue mathematician weeping.

that homeopathy is an effective treatment given the evidence from meta-analysis studies showing it to be no more effective than placebos in well administered studies? What is the probability that aliens have visited earth given the great distances involved and what we know from special relativity? While wording our inquires like this may not give us a definitive answer it should at least point us in the right direction, and it is certainly better than asking what is possible.

Science! It Works, Bitches!

You cannae change the laws of physics Jim
Energy, entropy and quantum mechanics

We have now come to the point when we need to take a look at some of the more complex area of modern science. Now I am not even going to pretend that this will be an in depth look at such subjects as thermodynamics and quantum mechanics, partly because I am far from an expert on these subjects myself and partly because even a quick overview of one of these topics would easily fill the rest of this book. So why mention them at all? Well as I mentioned at the beginning of the book these are areas of science that proponents of the paranormal and pseudoscience like to use to support their claims, mainly due to the fact that so few people know enough about these mind boggling issues to adequately refute their arguments. As such, having a basic understanding of what these subjects are about, and more importantly what they are not about, will allow you to see through the complex terminology and high minded ideas used by proponents of pseudoscience to obfuscate the nonsense hiding behind beneath.

Energy

No end of paranormal and pseudoscientific claims talk about energy in some form or another. Whether it is the life energy of chiropractic, the chi energy of acupuncture or the strange, supernatural energy of auras, orbs and ley lines, the World of the paranormal and pseudoscience is all about energy. The thing is

Science! It Works, Bitches!

that in all of these cases the proponents are demonstrating just one thing. That they have no idea what energy actually is, and if we are going to talk about thermodynamics and quantum mechanics then knowing what energy actually is is a good place to start.

So what is energy? Well if you listen to the proponents of the supernatural then you would be forgiven for thinking that energy is an actual thing, some glowing, semi translucent, hovering ball of every changing colour from which power can be drawn to perform everything from psychic healings and communication with the dead to spoon bending and amazing feats of mind over body, such as those made famous by the Shaolin monks. Energy is also often described in terms of health. If we are feeling a bit under the weather then, the proponents claim, it is due to a problem with our energy field. In order to be returned to good health our energy needs to be "rebalanced", "unblocked", "tuned", "aligned" or in some other way manipulated to get it to once more flow as it should. But don't take my word for it. I did a quick search on Google for "energy fields" and the very first page to pop up was an article entitled "Auras, Chakras and Energy Fields: Cleansing and Activating Your Energy Systems" by Mary Kurus, a Canadian "Vibrational Consultant." Here's what she has to say on this subject:

> "Our energy systems are alive and intelligent. They know exactly what they need for perfect health and vibrant energy. When our energy systems are disrupted, blocked, slowed down, or damaged, messages are sent to the conscious level that something is wrong and that we need to address imbalances, blockages and damage. We are trained to look at the physical body and address its needs rather than look at our energetic systems to provide them what they need for

perfect health. Our energy systems are conscious energy and this energy is connected with all energies in the Universe. Our energy systems act as a connection between the physical world and the metaphysical or spiritual world. When we look after our energy systems we look after our deepest and most profound needs, including the physical, emotional intellectual and spiritual."

Sounds impressive doesn't it, but does it make any sense? Well, no, no it doesn't, and that's because of what energy really is. You see energy isn't really a thing at all and should really not be used as a noun in the way people like Mary use it. Energy is actually a form of measurement, just like the centimetre or mile. In this case it is a measurement of the ability for something to do work. A very easy way to understand this is with regards to something we have already talked about, and that's Einstein's famous $E=MC^2$ equation. Now as you know this equation explains the interaction between energy and mass, that being the amount of energy something has is equal to its mass multiplied by the speed of light squared. Now mass is basically a measurement of weight and the speed of light is, unsurprisingly, a measurement of the speed at which light travels. These two measurements, combined in Einstein's beautiful equation, results in a measurement of that object's ability to do work. For example the packet of beef flavoured Monster Munch crisps I am currently eating contain 451kj calories of chemical energy that, hopefully, will enable me to carry out the work required to complete this chapter.

As well as chemical energy there are many other different types of energy, including kinetic, potential, thermal, sound, light, gravitational, and electromagnetic. Any form of energy can be transformed into any other form but, as the law of conservation of energy tells us, the total amount of energy always remains the same. The important thing to remember however with regards to this discussion is, that no matter what form of energy you are talking about, energy needs to be stored in some way or another. It simply doesn't exist completely independently on its own. When faced with someone talking about energy in relation to some New Age claim the question you should be asking is "how is the energy stored?" Are they talking about heat or maybe some sort of chemical energy such as that stored in petrol? Is their spiritual energy of a similar kind to that found in my packet of Monster Munch?

Simply understanding what energy is will stand you in good stead when it comes to evaluating the legitimacy of any claim presented to you that uses the term. As with so many other areas when it comes to skepticism I find myself indebted to Brian Dunning, creator of the Skeptoid podcast, and his ability to get to the heart of any skeptical matter, for the best advice in this area. Brian's simple but effective suggestion when faced with any claim regard some form of energy is to just replace the word energy in your mind with the phrase "measurable work capacity." If the sentence still makes sense then there is a good chance that the claim is using the word energy in the correct way and as such has a better chance of being scientifically accurate. With than in mind flip back and re-read the quote from Mary Kurus, only this time replace each use of the word energy with the phrase "measurable work capacity." Does it still sound impressive or make any kind of sense?

Free Energy

Now that we have a working definition of energy we can move on to take a look at the subject of thermodynamics. In an earlier chapter I quoted Dr Steven Novella regarding the fact that while we can never be 100% certain of anything in science there are things that have been confirmed to such a ridiculously high degree that to look at them as anything other than cold hard facts is just insane. The example of something that can be taken as a cold hard scientific fact that Dr Novella used was the first law of thermodynamics, itself an expression of the law of conservation of energy, which states that energy can neither be created nor destroyed only changed from one form to another. Simply put, with regards to energy, there is no such thing as a free lunch, you just don't get something for nothing. When it comes to science things don't get much more certain than this.

As I said this principle of thermodynamics comes from the law of conservation of energy, which states that the total amount of energy in an isolated system never changes, it always remains constant. Imagine a car with a full tank of petrol. The petrol contains a set amount of energy that is stored in the form of chemical energy. When you start the car and drive away petrol is fed into the engine where that chemical energy is converted into the kinetic energy that moves the car as well as the heat that comes off the engine. When you brake the kinetic energy of the car is converted to heat energy in the brakes. In reality it is a little more complicated than that but the end result is that the total amount of energy that started out in the petrol remains the same even though it has now been converted to a number of different forms such as kinetic energy and heat. Let me give another, slightly clearer example.

This time let's imagine a light bulb. A light bulb converts electrical energy into light energy. If we were to measure the amount of energy involved on each end of this equation at a given moment then we might see something like this:

Electrical Energy in	*Converted to*	**Light Energy out**
100 J (joules)		10 J (joules)

Now at first glance this may appear to invalidate the claim that the total amount of energy always remains the same, I mean where the hell has the other 90 joules of energy disappeared to? Well, as anyone who has ever made the mistake of trying to change a light bulb just after it has blown will know, that extra energy doesn't just disappear, it gets converted to heat. In fact your average light bulb only converts about 5-10% of the electrical energy into light energy, the rest gets turned into heat. With that in mind what our little equation really should look like is this:

Electrical Energy in *Converted to* **Light Energy out**
100 J (joules) 10 J (joules)
and
Heat Energy out
90 J (joules)

Once again the total amount of energy involved remains the same at the end of the process as it was at the beginning, only it has been divided into a number of different forms. It's worth noting that in both the examples given some of the energy involved has been converted into heat and this is one of the most common by-products of converting energy from one form to another. With the car the goal is to convert all of the chemical energy in the petrol into kinetic energy to move the car and with the light bulb the aim is to convert all the electrical energy into light energy. But in both cases some of the initial energy is converted to heat as well and as this is not our goal this heat energy can be viewed as "lost" energy. This holds true for any form of energy conversion.

Building your Skeptical Toolkit

Whenever you convert one form of energy into another some energy is always lost in the process, usually as heat. The difference between the amount of energy put in and the amount of useful energy you get out of a system tells you how efficient that system is. In the previous example, the light bulb is only 10% efficient. Hmm might be time to switch over to those energy saving bulbs.

And that's actually pretty much it with regards to the first law and the conservation of energy, aside from a whole load of really complex mathematics, and so apart from being something that everyone really should know how does this tie back to skepticism? Well this universal scientific principle is the reason why we can dismiss immediately anyone's claim to have created a perpetual motion or free energy device. Such devices violate the laws of thermodynamics, and not just the first one, and as such simply cannot work the way the proponents claim.

While there are many hundreds of different designs when it comes down to it there are just two basic types of perpetual motion devices, those that violate the first law of thermodynamics and those that violate the second law of thermodynamics, and we'll come to that second type in a moment. While the term perpetual motion implies a device that works forever it is more commonly used to describe machines that produce more energy than is put into them. As such, what people tend to be talking about with regards to perpetual motion devices are actually "free" energy devices.

As we know this violates the law of conservation of energy. You can't get more energy out of something than you put it. It also goes against the fact that when converting energy from one form to another there is always a net loss of useful energy. Generally if someone claims that they are getting more energy out of their device than they are putting in then they are not accounting for all possible sources of energy, that or they are just lying to you.

Entropy and the 2nd Law

Entropy is another one of those terms that is often misused by proponents of pseudoscience. A really quick and simple, though not completely accurate, way to think about entropy is as the opposite of energy. As we covered earlier energy is the measurement of something's ability to do work. Entropy on the other hand is the measurement of the amount of energy in a system that is *not* available to do work. Let me put that another way. As we have already seen energy can neither be created nor destroyed, but it can be changed from one form to another. We have also covered that this exchange of energy is not perfect and that some energy is always lost, most often as heat, in the conversion process. Well sometimes this lost energy ends up in a form that is no longer available for doing work and entropy is a measure of this type of energy. Of course in actuality it is a lot more complicated than this and entropy, as with energy, is far more than just the definition of the term, however for our needs this simple explanation will suffice.*

Now the most common situation where the term entropy is misused is when anti-evolution proponents, often religious but not always, talk about the 2nd law of thermodynamics as a way of trying to disprove the validity of the theory of evolution. As an example here is a quote taken from an article entitled "Evolution, Thermodynamics, and Entropy" by Henry Morris, who had a Ph.D. in hydraulic engineering, was one of the founders of the Creation Research Society and the Institute for Creation Research and was

---

* And I am pretty sure I just made physicists weep with that comment.

generally considered to be "the father of modern creation science".

> "Remember this tendency from order to disorder applies to all real processes. Real processes include, of course, biological and geological processes, as well as chemical and physical processes. The interesting question is: "How does a real biological process, which goes from order to disorder, result in evolution. which goes from disorder to order?" Perhaps the evolutionist can ultimately find an answer to this question, but he at least should not ignore it, as most evolutionists do.
>
> Especially is such a question vital, when we are thinking of evolution as a growth process on the grand scale from atom to Adam and from particle to people. This represents in absolutely gigantic increase in order and complexity, and is clearly out of place altogether in the context of the Second Law."

To summarise Henry's argument, as evolution requires a massive increase in order over millions of years to work and the 2nd law states that things process from order to disorder, evolution must therefore be incorrect. Now at face value this seems to make sense, however the whole line of reasoning is based upon a misunderstanding of the 2nd law and of what entropy actually is. Before we go on however let's look at what the 2nd law of thermodynamics actually says:

> "The second law of thermodynamics is an expression of the universal law of increasing entropy, stating that the entropy of an isolated system which is not in equilibrium will tend to

increase over time, approaching a maximum value at equilibrium."

Yeah, that's going to need some clarifying. Basically the 2nd law describes the tendency of heat to flow from hot systems to cold systems homogenizing, or balancing out, the amount of heat in the system by making it available to the maximum number of different areas. Think of an ice cube in a warm room. The cold from the ice cube doesn't flow out into the warm room to cool it down, rather the heat from the room flows to the ice cube warming it up. Heat flows from hot areas to cooler areas and never the other way around[*]. The other thing that the 2nd law talks about is that the amount of entropy in an isolated, or closed, system always either increases or stays the same. Now this does not mean that there cannot be areas in which entropy decreases, only that there must be a larger increase in entropy elsewhere in the system to compensate. Water freezing into ice in a freezer represents a decrease in entropy; however the heat given out by the freezer in order to do this results in a net increase in entropy.

The mistake that Henry makes when talking about the 2nd law is that he equates entropy with disorder. Entropy is not a measure of disorder, despite often being stated as such, but rather a measure of the amount of energy in a system that is not available for doing work. As such the 2nd law is in no way talking about the "tendency from order to disorder." Because of this misunderstanding, it is easy to make the mistake that the 2nd law somehow disproves evolution which, in case you were confused, it most definitely does not. And even if entropy did mean disorder the Earth is not an isolated system as we get energy from the Sun, a process in which, fitting perfectly with the 2nd law, heat flows

---

[*] At least not naturally, your fridge happily transfers heat from cold to hot areas all day long.

from a hot area to cooler ones. And yes, while evolutionary processes do cause a decrease in entropy the vast increase in the entropy of the Sun results in a net increase in entropy of the entire system, which again more than fits with the 2nd law. Sorry Henry.

Which brings us back to the second type of perpetual motion devices. These are machines that claim to draw heat from the environment and convert it into energy. Now if the thermal energy involved is equal to the amount of work done then this does not violate either the first law or the conservation of energy so why is it that these kinds of devices are not possible either? Well it's because they do violate the 2nd law. When you transfer heat from a hotter area to a cooler one the latter is heated up in the process. It is not possible to have a device that draws heat from the environment and does not heat up in the process. This is exactly what a perpetual motion device of the second type claims to do and once again the laws of thermodynamics mean this method can't work either.

Quantum mechanics

Just the mention of quantum mechanics is enough to make most people's heads spin. Those who peddle new age treatments and pseudoscientific quick fixes rely on the fact that most of us don't know out Planck's constant from our wave-particle duality in order to get away with claiming that quantum mechanics supports their claim. Here's an example from Deepak Chopra, an Indian born doctor, celebrity guru and purveyor of something called "quantum healing."[*]

---

[*] Taken from an interview with Healthy.net conducted by Daniel Redwood D.C.

Science! It Works, Bitches!

> "Quantum healing is healing the bodymind from a quantum level. That means from a level which is not manifest at a sensory level. Our bodies ultimately are fields of information, intelligence and energy. Quantum healing involves a shift in the fields of energy information, so as to bring about a correction in an idea that has gone wrong. So quantum healing involves healing one mode of consciousness, mind, to bring about changes in another mode of consciousness, body."

Note that word energy being misused again there. But when it comes to the term quantum it is very understandable that people can be easily bamboozled, as quantum mechanics is a vastly complex subject. In fact Richard Feynman, winner of the Nobel Prize in physics, once famously stated, "If you think you understand quantum mechanics, you don't understand quantum mechanics". So, how is it then that I can confidently state that quantum mechanics does not in fact support ESP, dowsing, psychic healing and all manner of other psi phenomena if I openly admit the subject is beyond my grasp? Well, as always, a little knowledge can be a dangerous thing.

First off here is my two pence worth, really super basic intro to quantum mechanics. Simply put quantum mechanics is the study of matter and energy at the atomic level. The physical laws that govern macroscopic objects, such as ourselves, do not function in the same way once you get down to the microscopically small atomic realm. Whereas Newtonian physics describes the actions of things (apples, planets, balls, etc) on the macro level very precisely, quantum mechanics deals with the very small (molecules, atoms, electrons, protons, neutrons, etc) by primarily using probabilities.

Building your Skeptical Toolkit

That aside the quantum world is serious weird and the completely counter intuitive nature of quantum mechanics can be summed up fairly well by looking at the story of Schrödinger's cat. Now whether Erwin Schrödinger actually had a cat or not is beside the point[*] as this was never actually a real experiment but rather just a thought experiment.

You place a cat in a sealed box along with a glass container of some kind of poison and a source of radiation. A Geiger counter is also set up inside the box and if it detects radioactive decay then the container is broken, releasing the poison and killing the cat. Now applying what is known as the Copenhagen interpretation of quantum mechanics to this experiment it can be stated that after a while the cat in the box is simultaneously both alive and dead. Using the technical term the cat would be in a state of quantum superposition. Of course if we were to look in the box we would find that the cat was either alive or dead and not a mixture of the two.

Things at the quantum level however have the ability to be in multiple different states at the same time. For example light is both a particle and a wave at the same time. However, the act of measuring light, just like opening the box in the case of Schrödinger's cat, causes all of the various probable states to "collapse" to the value defined by the measurement. Hopefully you are still with me.

Now the point of Schrödinger's thought experiment, and the reason I bring it up here, was to show the problem of applying quantum mechanics to the macro world. While light can indeed be both a particle and a wave at the exact same time a cat cannot be both alive and dead at the same time. It is either one or the other and the fact that we may not know which simply doesn't come into the equation. At no point is the cat, or anything else on

---

[*] Though I imagine the RSPCA might have called round if he had.

137

the macro level, in a state where it is actually both alive and dead at the same time. Remember those laws of logic?

This is why we can pretty much rule out quantum mechanics as a cause for anything that operates on a macro level, such as healing someone with the power of crystals. All paranormal and pseudoscientific claims involving quantum mechanics aim to produce results on a macro level, otherwise how would we tell if they were working or not? Now it may be true the psi powers can alter things on the quantum level, we currently really can't tell for sure, but seeing that the physics that govern things on this scale simply don't apply to the macro world, the World that we as human beings interact with and care about, then doing so would not have, and could not have, any measurable effect on our well-being. At the end of the day in the World of pseudoscience the term quantum mechanics, along with the word energy, is so incorrectly used that, though it may sound high tech and impressive, it is ultimately meaningless.

Time for a little Mad Science
How to Identify pseudoscience

Ok so we have spent some time going over what makes up good science and how it works, but for skeptics this is only half the story. As well us understanding good science you also need to be able to spot bad science, or pseudoscience, when you encounter it. Well I'm sorry to say that this is sometimes a lot easier said than done. Sadly there is no one thing, no flashing warning sign, which clearly identifies something as a pseudoscience, but that's not to say that there aren't red flags for you to be on the lookout for. Many skeptical authors over the years have produced lists of just these sort of things, the presence of which will indicate that you might be dealing with a pseudoscience, and I've gone through as many of them as I can find and have combined the wisdom of these great skeptical thinkers into what I hope is a pretty definitive list of pseudoscientific red flags.

Remember not every pseudoscientific claim you come across will display all of the red flags on this list and sometimes even legitimate scientific claims might be flying one or two. However, the more items on this list you can tick off the more likely you are to be dealing with a pseudoscientific idea or claim. There may not be a flashing warning sign attached to every pseudoscience but once you know what to look for you should have little trouble telling the difference between the real McCoy and the pale imitation.

Is it supported by good quality evidence?

The first and most important thing that separates real science from pseudoscience is good quality evidence. In science evidence forms the foundation upon which all else is built. Scientific theories stand and fall on the quality of the evidence supporting them. Good science is built upon experimental results and statistical data that, and this is the important part, must be independently verifiable. The supporting evidence should come from a variety of sources and claims should be based upon a myriad of supporting columns rather than a single foundation. Evidence should be accepted or rejected based upon how reliable, accurate and, where possible, reproducible it is and never because it does or does not support the desired conclusion. The evidence should also be directly relevant to the theory at hand. Many pseudoscientific claims like to build themselves up by referencing legitimate science that is in reality only tangentially related to the claim they are making. Where ever possible you should always search out your own evidence and not just rely on what is presented to you by the proponents of the claim. It is always possible that they have cherry picked the evidence in order to present you with the strongest case possible and in hope that you won't discover the mountain of evidence against them on your own.

Can it be falsified/is it testable?

Just as important as good evidence real scientific claims should always be testable and, more importantly, falsifiable. If it is not possible to test, and thereby falsify, a claim then it is not a legitimate scientific theory. To pick a commonly used example it is often said that the discovery of a rabbit fossil in pre-Cambrian rock would falsify many of the claims of evolution. Now while this is unlikely to happen it is still a legitimate falsifier, as should it happen it would show that our current understanding of how life on earth evolved was incorrect.

Most pseudoscience on the other hand it not falsifiable and, it is sometimes claimed, not even testable. It is not uncommon for the proponents of pseudoscientific claims to state that it doesn't work under test conditions or, if it is tested and fails, to rationalise this away by blaming the test conditions or even the skeptical nature of the testers themselves. For example, a 2005 paper published in the Journal of Alternative and Complementary Medicine when so far as to label randomized clinical trials "redundant" because they consistently fail to prove the efficacy of homeopathic treatments. The author concluded that the randomized clinical trial process itself was at fault rather than acknowledge the overwhelming evidence showing that homeopathy just doesn't work. Because of this kind of thinking pseudoscientific claims will also tend to avoid the peer review process as this is specifically designed to subject testing methods to scrutiny and to encourage others to carry out the same tests for themselves. Alternatively, as is becoming more common, they may decide to produce their own journals in which to publish, thus allowing them to get the work out there but avoiding the rigorous peer review process that weeds out bad research.

Does it act like a real science?

Genuine scientific claims tend to act in a number of predictable ways that pseudoscientific claims don't. Real science progresses over the years as new evidence comes to light that either reinforces or negates current thinking. For example, while the fundamentals of the theory of evolution have changed little in the

Science! It Works, Bitches!

last one hundred and fifty years many of the details have. When Darwin first proposed the idea of natural selection it was not known exactly how traits were passed on, but once the discovery of DNA was made it provided an explanation for how heredity worked and so was integrated into the theory, thus changing and reinforcing it.

Real science also makes predictions about what we should find if the science is accurate. For example, astronomers were able to predict the existence of the planet Neptune due to the fact that the orbit of the planet Uranus did not fit with Newton's law of gravitation. Tellingly astrologers failed to make the same prediction despite claims that their system for divining information about people is supposedly based upon the position of celestial bodies. If there was really anything to this then astrologers, just like the astronomers, should have noticed errors in their divinations that pointed to the existence of another planet.[*]

Lastly all real scientific theories operate within limits and under set conditions that specify when the do and when they don't apply. The 2nd law of thermodynamics only applies to closed systems and many of the laws of physics start to fall down once you get back to the very first moments of the Big Bang. In medicine it is understood that specific treatments and drugs are used in specific situations, however the same does not seem to

---

[*] It is also worth noting here that since the signs of the zodiac used by astrologers were first established the time of year in which the sun is found in each one has changed, though astrology charts have not changed to reflect this. For example when the system was first established the sun crossed through the zodiac sign Aquarius between January 20th and February 19th. Now however it crosses through it between February 16th and March 11th. Astrologers argue that the signs of the zodiac are "fixed" and are therefore unaffected by such things, though this really only holds true for the "tropical zodiac" system. I will leave it up to you to decide if this constitutes another example of astrology not acting like science.

apply to many pseudoscientific modalities that often claim to be effective for pretty much everything that might be wrong with you. As for limits many forms of pseudoscience are said to work under all conditions, except test conditions that is.

Is it self-correcting?

This is another aspect of what makes science science, however I think it is important enough to warrant a separate heading. Because of the way it focuses on evidence, builds upon the work of others and the way that the peer review process actually encourages others to take the results of an experiment and try to prove them wrong, science is a self-correcting process. What this means is that if mistakes are made or someone tries to fake their results it will be discovered and corrected. Let's look at a famous example; Piltdown man. In 1912 what appeared to be the fossilized skull and jaw bone of an early hominid was "discovered" in Piltdown, East Sussex in England. This find was exactly what palaeontologists of the time had been looking for and the fact that it was found in England seemed to support the expectation that modern man may have originated in Europe. However, Piltdown man was challenged almost from the outset and as early as 1913 papers were being published stating that this find was actually an ape jaw and a human skull. Furthermore, Piltdown man just didn't fit with the other hominid fossils being discovered around the globe; most notably it did not fit with the wealth of fossils that came out of Africa in the 1930s. Finally in 1953, over forty years since it had first appeared on the scientific scene, Piltdown man was officially ruled a deliberate fraud created by combining a modern human skull from the middle ages, the jaw bone of a 500 year old orangutan and fossilized chimpanzee teeth. The self-correcting nature of science had taken a "find" that seemed to support everything that thinking at the time believed about human evolution and had correctly identified it as a fraud because it did not fit with the overwhelming evidence against it. If an idea refuses to change in the face of good evidence against it then that is a very good indicator that you are dealing with pseudoscience rather than the real thing.

Does it fly in the face of known science or current thinking?

Now it is true that sometimes an idea comes along that seems to fly in the face of current scientific thinking but which ultimately turns out to be correct. The theory of evolution springs to mind, as does relativity and quantum mechanics. However, these events are rare and even with the examples mentioned these ideas did not appear in a vacuum and were more a different interpretation of the evidence rather than a complete contradiction of it. The same does not hold true for pseudoscience however. Here we are not just talking about a reinterpretation of the evidence but more an abject defiance of it. Let's take my favourite bit of quack medicine as an example.

One of the principles behind how homeopathy is said to work is known as the Law of Infinitesimals. Very basically this states that the more diluted a substance is the more effective it is. Now straight away this should strike you as odd, but let's stick with the science shall we. When homeopaths talk about diluting a substance they don't just mean dissolving a tablet in a glass of water before drinking it down. No they mean really dilute, really really dilute. German physician Samuel Hahnemann, the guy who came up with homeopathy in the first place, recommended that for most homeopathic remedies the active ingredient should be prepared to a dilution of 30C, which is the homeopathic way of saying a dilution of 1 in 10^{60}. Now in case you don't know what this means let me spell it out clearly. A 30C dilution means that there is one molecule of active ingredient for every 1 followed by 60 zeroes molecules of water. In case you can't picture what that looks like, but mainly because I find it so amusing, this is what 10^{60} looks like.

1,000,000,000,000,000,000,000, 000,000,000, 000,000,000,000,000,000,000,000,000,000

What this means is that in order to be sure that you were getting a single molecule of active ingredient in your homeopathic remedy you would need to consume 10^{41} homeopathic pills, which adds up to around a billion times the mass of the Earth. I hope you didn't have a big dinner. And some homeopathic remedies come at a dilution of 200C, or 1 in 10^{400}, which is a number in excess of the number of molecules in the entire universe. But all this is academic as it ignores a scientific principle known as the Avogadro constant.

The Avogadro constant is the point at which it is no longer possible to dilute something without losing the original ingredient altogether. Beyond this point you are just dealing with something that is chemically identical to pure water. The value of the Avogadro constant, which has been confirmed experimentally, is 6.023 x 10^{23}, which roughly corresponds to a homeopathic dilution of 12C, or 1 in 10^{24}. This means that for any homeopathic remedy with a dilution over 12C it is scientifically impossible for it to contain even a single molecule of the original active ingredient. Now Samuel Hahnemann can be forgiven for not knowing this, as the Avogadro constant was not discovered until after he founded homeopathy, however modern homeopaths have no such excuse, which may be why they cling so tightly to the equally discredited

Science! It Works, Bitches!

idea, proposed by French immunologist Jacques Benveniste in 1988, that water has a memory.

Does it pass Occam's Razor and Hume's Maxim?

Can the claim be explained more effectively using natural causes? Does it require you to assume things that are themselves not in evidence? If the testimony of the person telling you the claim was false would that be a more amazing and fantastical situation that the one the claim is describing? In the next part of this book we will look at a number of ways that the human mind likes to play tricks on us, for example how the law of large numbers allows for one in a million events to occur every day, and we have already seen how the placebo effect can actually increase our bodies natural ability to heal. With these things in mind is it more likely that the claim is accurate or that the person making it is misinformed, mistaken or just plain wrong?

Is it the invention of just one person?

While there are areas of science that can be said to be the creation of a single person, such as Newton's laws of gravity or Einstein's special relativity, even these amazing breakthroughs did not happen in complete isolation and built upon the work of those who came before them. Science tends to be a team sport, with lots of people adding to the work of lots of other people. The same does not hold true for many pseudoscientific claims however. Homeopathy for example was entirely the creation of German physician Samuel Hahnemann in 1796 while D.D. Palmer founded chiropractic almost completely on his own, with a little help from his son, in the 1890s.

Was it created before the invention of modern science/medicine?

This brings us nicely onto the next thing to be on the lookout for. Prior to the end of the 18th century very little was actually known about the cause of disease and almost nothing of what we now recognise as modern medicine existed. Before this there really was no such thing as the germ theory of disease and most of Western medicine was based upon herbalism and the ancient Greek idea of the "four humours". Now this does not mean that claims that come from a time before this are necessarily incorrect, however the fact that, for example, homeopathy was created before we really knew about the existence of germs should at least raise a red flag.

Is it said to be based upon ancient knowledge?

Again this point is related to the previous one. For some reason a lot of people are under the impression that a vast amount of scientific and medical knowledge comes from the ancient past. Many pseudoscientific claims rely on this misunderstanding and so claim to be based upon ancient knowledge. After all if the ancient Greeks or Chinese believed it to be true then it must be the case, right? Well actually this is not the case at all. In reality the more recent the discovery the more likely it is to be scientifically valid[*]. Science works by discarding bad evidence and building on the good evidence. Old theories are replaced by new ones as our knowledge grows. Just because something was

---

[*] Obvious exceptions apply to this line of reasoning and it would actually be a fallacy to assume that just because something is new it must be better.

common practice a thousand years ago does not mean that it is better or that we should prefer it over medical practices developed in the last few decades. Again this doesn't mean that everything from the past should be dismissed, but being old is not the same thing as being right...despite what your parents may tell you.

Does it claim to be "all natural"?

Everywhere you look these days you will see the claim that this or that product is "all natural". People mistakenly equate the term "all natural" with "all good", assuming that if something is natural then it must mean it is therefore safe and healthy for you. This is of course nonsense. There are many, many things that qualify as "all natural" that will kill you in no time. Hemlock, asbestos, E. coli, bubonic plague, smallpox, lead, box jellyfish neurotoxin, I could go on all day. Also in the vast majority of cases where you have a synthetic version of a naturally occurring compound they have been specifically engineered to make them safer and more effective. You could, should you be so inclined, boil up some willow tree bark and drink it down as a treatment for your headache. Or you could take an aspirin tablet instead, aspirin being the ingredient in willow tree bark that treats your headache, a tablet that contains only what is needed to treat your pain and none of the other stuff found in tree bark, such as wood. Natural doesn't always mean better.

Is it based upon an unknown form of energy?

As we learnt earlier in the book energy is actually a unit of measurement and unless you specify what sort of energy you are talking about and how you intend to measure it then the term is pretty much meaningless. But it sure sounds scientific and this is why promoters of pseudoscience love to use it. As such if someone starts talking about energy, and especially if they mention terms like Qi, energy fields, negative energy, spiritual energy, energy flow, auras, vibrations or vitalism, then this should raise a big red flag.

Does it use meaningless terminology?

In a similar way to how they use the word energy many pseudoscientific and paranormal claims like to use scientific sounding terms in order to give them the illusion of legitimacy. Unlike real science however these terms are often utterly meaningless and are used to confuse rather than clarify the situation. Something might be said to "[reverse] polarity in the energy flow" or cause "perturbations in the thought field"[*] but these claims fail to actually explain anything. And of course then there is the quote from alternative medicine guru Deepak Chopra that I used in the previous chapter:

> "Quantum healing is healing the bodymind from a quantum level. That means from a level which is not manifest at a sensory level. Our bodies ultimately are fields of information, intelligence and energy. Quantum healing involves a shift in the fields of energy information, so as to bring about a correction in an idea that has gone wrong. So quantum healing involves healing one mode of consciousness, mind, to bring about changes in another mode of consciousness, body."

---

[*] Both of these examples are actually used by Dr. Roger Callahan to describe "Thought Field Therapy".

Science! It Works, Bitches!

In science the point of using specific and often complex terms is to bring clarity not sow confusion. Talk of quarks, leptons and neutrinos for example may sound as confusing as Deepak Chopra's statement above, but these terms have very specific meanings and any scientist with knowledge of what these words mean would understand exactly what any other scientist was talking about if they used them. If you wanted you could study up on these terms and then you too would be able to understand exactly what particle physicists mean when they talk about quarks having up and down spin. Conversely can anyone say for sure exactly what Chopra means when he says "information" or "energy" or talks about "healing one mode of consciousness"?

Does it claim to be based upon quantum physics?

As you can see from the example above Deepak Chopra really likes to recourse to quantum physics and he is not alone in this. Quantum physics is a highly complicated subject and deals with all manner of strange and counter intuitive things taking place on the quantum level, a level that is completely alien to the World in which we live on a daily basis. As such many proponents of pseudoscience like to use it as an explanation for the claims they are making. As with the use of the word energy and other scientific sounding terms it is used in the hope that it will give the claim a level of legitimacy it does not have and to prevent the layman from enquiring further. Any mention of quantum physics, especially in relation to any medical claim, should set alarm bells ringing.

Does it rely on appeals to history, public opinion or common sense?

Real science doesn't care about how things were done in the past, what the public opinion on the matter is or whether or not the conclusions fit with common sense. Real science cares about what the evidence shows and if the evidence goes against the conclusions of the general public well then so be it. Pseudoscience however tends to fly in the face of the evidence and as such is more than happy to rely on these things. As such if you come across statements like those listed below being used in defence of a claim then it should send up a red flag indicating that you are not dealing with real science.

> "90% of people believe…"
> "Everyone knows that…"
> "People have used this since the 15th century."
> "Well is just makes sense that…"

Common sense in particular can be a dangerous thing. Common sense once told mankind that the earth was flat, that the sun orbited around us and that there was no way we could ever walk on the moon. Science tends to laugh in the face of common sense. If all someone has in support of their claim is a recourse to common sense then that is a very good reason to be skeptical.

Was the claim first announced to the press or through popular media?

As we saw earlier the peer review process is an important part of the scientific method. It is there to help identify mistakes and to

Science! It Works, Bitches!

catch deliberate frauds. It is designed to make sure that what is being reported is good science supported by equally good evidence. It is a tough, often harrowing gauntlet in which a scientist may find their work rejected numerous times before it is finally published, only for it to be torn to bits by other scientists working in the field when they are unable to replicate the published results. Now this may seem like a lot of stress and hard work, and it is, but it also makes sure that the science that comes out at the other end is the good stuff and if a scientist has correctly followed the scientific method and has the evidence to back up their claims then they have nothing to fear from the peer review process. So when some people choose to skip this process entirely and announce their findings directly to the general public by way of the media you have good reason to be skeptical. The famous, and now completely discredited, Pons and Fleischman cold fusion experiments are perhaps the most well-known example of this, but I prefer to use a more modern example of bypassing the peer review process.

In August 2006 an Irish company by the name of Steorn took out a full page advert in The Economist magazine announcing that they had developed a device that produced "free, clean and constant energy". They were looking for twelve of the World's best scientists to come take a look at their device, which they later dubbed the Orbo, and verify their claims. Almost a year later they were once again back in the news announcing that they would be giving a public demonstration of the device that could be viewed live on the internet. Strangely they experienced "technical difficulties" and the demo was called off. Finally in July 2009 it was announced that an independent jury of scientists had been unable to show that the Orbo device actually produced energy, which didn't exactly come as a surprise to anyone with even a basic understanding of the laws of thermodynamics. If they had just gone through the peer review process in the first place they could have saved themselves all the public embarrassment. Orbo, as with any free energy device, is pure pseudoscience.*

Does it rely on personal testimonies as evidence?

As we saw earlier anecdotal evidence is perhaps the weakest form of evidence when it comes to evaluating the legitimacy of scientific claims. And yet it is one of the most popular forms of evidence used in support of pseudoscientific claims. While websites promoting legitimate science will most likely quote peer reviewed research and experimental data in order to back up their claims, sites promoting pseudoscience much prefer personal testimonies, relying on the subjective feelings of their customers, already a self-selected group, rather than the objective reality of double blind clinical trials.

Does it rely upon corny marketing or the personality of the person promoting it?

Because they often don't actually have much evidence to support their claims promoters of pseudoscience tend to prefer to rely on marketing and celebrity endorsement to give their claims the illusion of legitimacy. A worrying number of celebrities are more than happy to lend their support to all kinds of pseudoscience. Actress Jenny McCarthy has become the face of the anti-vaccination movement in the United States and Oprah Winfrey

---

* Interestingly as I write this in August 2011, almost five years after it was first announce, Steorn are still working on the Orbo device with no apparent sign that they view the laws of physics as a barrier to their eventual success.

Science! It Works, Bitches!

has lent her considerable influence[*] and the support of her vast media empire to everything from angels to vitamin mega dosing, acupuncture to psychic powers. Closer to home Prince Charles is a well-known supporter of all manner of alternative medical treatments including endorsing homeopathic "vaccines" for polio, typhoid and yellow fever, all dangerous and potentially fatal diseases. Now while celebrities do often lend their fame to legitimate causes and concerns as well, I think it is rather telling that you never see celebrity endorsement for things like Aspirin, it's simply not needed as the evidence speaks for itself.

As for the marketing side of things, big advertising campaigns are fairly easy to spot and I am sure all of us are a little bit skeptical about any form of advertising we encounter. What I am talking about here is a subtler approach that is again designed to give the illusion of legitimate science. Pictures of people in white lab coats doing "science" with test tubes and Bunsen burners in front of black boards covered in equations. Organisations that have the word "Institute" or "Academy" in their title. Websites and documentation that are full of personal testimonials and anecdotes rather than actual evidence. And then you have those impressive sounding qualifications that are often not worth the paper they are written on.

In the UK Dr Gillian McKeith PhD is a well-known celebrity nutritionist and has sold millions of copies of her book *You Are What You Eat*. Members of Scottish political parties have sought to have her advise the government and the Soil Association, a UK based charity organisation promoting the use of organic products, has given her a Consumer Education Award. The problem is that when you look into her qualifications they are less than impressive, as is her understanding of the science her

---

[*] Oprah is generally considered to be one of, if not the, most influential woman in the world. A spot on one of her shows can launch a career.

Building your Skeptical Toolkit

qualifications are supposedly in. In 2007 a successful complaint was lodged against her with the Advertising Standards Agency over her use of the term "doctor" in her advertising, on the grounds that her PhD came from a non-accredited American correspondence college. This resulted in the ASA telling her she was no longer allowed to claim to be a doctor in the advert in question, as it allegedly breached two clauses of the Committee of Advertising Practice code, namely those pertaining to "substantiation" and "truthfulness". Now does this mean that she doesn't actually have a PhD? Dr Ben Goldacre, an actual medical doctor, would argue that this is exactly what it means and in 2010 he looked into the matter further, publishing his findings in his column in The Guardian newspaper.

> "If you contact the Australasian College of Health Sciences (Portland, US) where McKeith has a "pending diploma in herbal medicine", they say they can't tell you anything about their students. When you contact Clayton College of Natural Health to ask where you can read her PhD, they say you can't. What kind of organisations are these? If I said I had a PhD from Cambridge, US or UK (I have neither), it would only take you a day to find it."

But even if her PhD is a little suspect McKeith is still a certified professional member of the American Association of Nutritional Consultants, though seeing that Goldacre's dead cat Hettie is also a certified professional member this is again a less than impressive claim. These days it is very easy to get official sounding qualifications in almost any subject for very little effort, and often a small payment, over the internet. Heck you have probably been sent emails offering you a PhD in the subject of your choice yourself. There are also lots of unaccredited colleges out there

Science! It Works, Bitches!

and as such, where ever possible, you should check out the legitimacy of the qualifications being presented to you, especially if they are given as evidence in support of a given claim. I mean I myself have a doctorate of science in parapsychology, I have a certificate and everything, and it only took me 30 seconds to get it.[*]

Does it have support that is political, ideological or cultural in nature or come from a source dedicated to supporting it?

The way good science works is to start with the evidence and follow it where ever it may lead in order to reach a conclusion. Pseudoscience very often operates the other way around. They start with a conclusion, often based upon some moral, ethical or political belief they have about the World, and then they go looking for evidence to support this conclusion. Sometimes entire organisations are formed to support these beliefs and they will often try to convince people than not only are their ideas right, but that it is the moral thing to accept their ideas as well. Organisations like this generally tend to act in a very biased way, reporting only on evidence that supports their ideas or goes against the prevailing scientific alternative. Evidence that goes against their existing conclusion is downplayed or just ignored outright. One of the clearest and best examples of this can be found in the Statement of Faith of Answers in Genesis, a US based Christian apologetics organisation primarily focused on promoting a Biblical view of the universe's origins, part of which states:

---

[*] You can get one yourself if you would like, just head over to thunderwoodcollege.com.

Building your Skeptical Toolkit

"By definition, no apparent, perceived or claimed evidence in any field, including history and chronology, can be valid if it contradicts the scriptural record."

If the evidence doesn't agree with you well then just ignore it, hardly a scientific approach. Often the organisations dedicated to promoting these pseudoscientific ideas will, rather than going through the peer review process, target public opinion directly, either with flashy campaigns, celebrity endorsement and clever marketing, as discussed in the previous point, or through the court systems. In America, much more than in the UK, supporters of Biblical creationism, like Answers in Genesis and other young Earth creationist organisations[*], have regularly tried to use the US court system to force schools to teach Biblical creationism, and the most recent iteration Intelligent Design, alongside or in place of subjects like evolution in science classes. For example the Discovery Institute, a Seattle based religious think tank, has a "Teach the Controversy" campaign which both promotes intelligent design creationism and tries to get the teaching of evolution removed from public high schools in the United States and which has already resulted in a number of high profile and expensive law suits. The campaign plays on the idea of fairness, stating that if evolution is being taught in schools it is only fair to teach the alternative point of view as well. At face value this sounds reasonable, but this is not how good science works. If

---

[*] In case you are unaware Young Earth Creationists believe that the planet and by extension the entire universe is only around 6000 years old. This age is reached by looking at the genealogies and ages of the various people mentioned in the Bible and working backwards to Adam, Biblically the first man created by God. Using this method Bishop James Ussher calculated in 1658AD that the Earth was created on October 23rd 4004BC and this is generally the date and age used by most Young Earth Creationists to this day. Some, both believers and skeptics, choose to celebrate the Earth's birthday on this day, though for different reasons.

Science! It Works, Bitches!

something is supported by the evidence and has made its way through the peer review process to be accepted by the scientific main stream then it will ultimately end up being taught in schools around the World, even if it takes a while to get this far. Trying to bypass this process and go straight to the schools is a sure sign of pseudoscience at work. Just think of all the other ideas we would have to teach if the "Teach the Controversy" idea was applied to other areas of science?

Some of the great "Teach the Controversy" T-shirts designed by Amorphia Apparel as a response to the Discovery Institute's campaign – controversy.wearscience.com

Does it make vague claims?

You may be unaware of this but in many countries it is actually illegal to make specific medical claims if you do not have the evidence to back them up. As such many forms of pseudoscience use vague, medical sounding phrases that don't actually make any specific claims but give the illusion that they do. While a box of Aspirin will list the specific conditions it treats, as well as possible side effects, a pseudoscientific treatment will more likely claim to "promote wellness" or "boost the immune system". These things sound scientific but don't actually give any useful information what so ever. Alternatively it may make no medical claims at all and simply rely upon the purchaser's assumption that it actually does something. Recently my Mother purchased some magnetic therapy plasters, at the recommendation of a member of staff at the chemists, to help treat the pain in my Father's arthritic knees. Nowhere on the box or the accompanying paperwork does it claim to treat, well, anything at all. Now does this mean that it doesn't treat arthritis? Well no of course not, and neither does the fact that research into other similar magnetic therapy devices which specifically shows that they are completely ineffective at treating arthritis mean that this particular product is likewise ineffective, and it would be wrong to jump to this conclusion. These plasters could indeed be great at treating arthritic pain, it is a possibility that a good skeptic should accept, however the fact that the packaging makes no medical claims should at least raise a red flag if nothing else.

Does it just sound too good to be true?

As the saying goes if something sounds too good to be true then it probably is. In real life there is rarely a single simple answer to a complex array of problems. When was the last time a claim about unlimited energy actually turned out to be true or a single pill that would make you thin again did just that? As Carl Sagan said "extraordinary claims require extraordinary evidence." Any product or service that claims to be world changing should be approached with severe skepticism.

Science! It Works, Bitches!

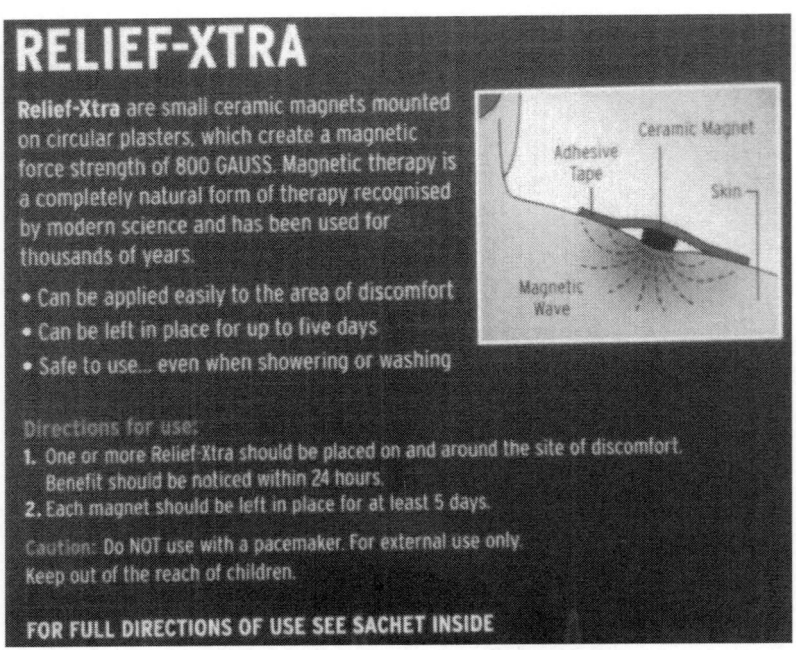

RELIEF-XTRA

Relief-Xtra are small ceramic magnets mounted on circular plasters, which create a magnetic force strength of 800 GAUSS. Magnetic therapy is a completely natural form of therapy recognised by modern science and has been used for thousands of years.

- Can be applied easily to the area of discomfort
- Can be left in place for up to five days
- Safe to use... even when showering or washing

Directions for use:
1. One or more Relief-Xtra should be placed on and around the site of discomfort. Benefit should be noticed within 24 hours.
2. Each magnet should be left in place for at least 5 days.

Caution: Do NOT use with a pacemaker. For external use only. Keep out of the reach of children.

FOR FULL DIRECTIONS OF USE SEE SACHET INSIDE

This product makes no claim to actually treat or reduce discomfort. The benefits it mentions are not specifically claimed to come from the use of this product, even if the implication is there. A sure sign that you should be skeptical.

Does it claim to treat a wide range of unrelated condition or act as a cure all?

Related to the previous point is the claim that treatment X will cure all that ails you. You've probably all seen films where a horse and cart rolls into town loaded down with a magical tonic that will cure you of all your ills. Maybe you even sang the song Lily the Pink as a child about Lydia Pinkham and her amazing "women's tonic" that was said to be "most efficacious in every case". Well the modern day equivalents to these people still hawk their wares to this day. Chiropractic for example may well be effective for

Building your Skeptical Toolkit

treating back pain but many chiropractors also claim that it treats such conditions as ADHD/learning disabilities, high blood pressure, baby colic, asthma, irritable bowel syndrome (ISB), bedwetting, menstrual cramps and dizziness,[*] things for which there is "not a jot of evidence".[†] It is also worth mentioning that anything that claims to cure "cancer" should be viewed skeptically. Though often thought of as a single condition the term "cancer" actually covers a wide range of different medical conditions. As such, an effective treatment for one form of cancer may be ineffective or even harmful if applied to another form. Therefore, it is highly unlikely that there will ever be a single cure for "cancer" as a whole.

Do the proponents claim that it is being suppressed by the authorities?

Convicted fraud and purveyor of pseudoscience Kevin Trudeau[‡] wrote a series of books all of which had titles that were some variation of "Natural Cures "They" Don't Want You To Know About" in which he claimed that cures for medical conditions

---

[*] Or at least they did. Here in the UK many Chiropractors no longer make these claims, at least in their literature and on their websites, after the McTimoney Chiropractic Association advised its members to remove all such claims following numerous complaints raised with the Trading Standards Institute in 2009 about these claims. These complaints came following the Advertising Standards Authorities ruling that Chiropractors could not claim to treat IBS, colic or learning difficulties due to a lack of scientific evidence that they could actually do so.

[†] To quote heroic skeptic Simon Singh from the Guardian article on Chiropractic that saw him wrongfully sued for libel by the British Chiropractic Association. The BCA subsequently dropped the case in 2010 when the court ruled that Simon's comments constituted "fair comment."

[‡] And yes I am aware I just engaged in the poisoning the well fallacy just then, but strangely I don't feel too bad about it ☺

ranging from cancer, herpes, and AIDS through to diabetes, multiple sclerosis, lupus and chronic fatigue syndrome were all being covered up by various government agencies and pharmaceutical companies. This kind of thing is common when it comes to pseudoscientific claims and is generally used as a way to explain why mainstream science or medicine does not take them seriously. Proponents often claim that there is a conspiracy within the scientific and medical community to keep their wonder cures from the public. Why? Well, they claim, it's because the medical industry has an interest in keeping you sick. The fact that in reality any doctor or company that did produce a cure to even one of the conditions mentioned above would make a fortune apparently doesn't enter into their thinking. It also plays into the psychological condition of in-group thinking. You feel special as you are one of the people who knows about this important piece of information and as such you are less likely to question its legitimacy.

Does it make an appeal to ignorance or wonder?

> "There are mysteries all around us."
> "Science can't explain everything."
> "The universe is a strange and mysterious place."

Many forms of pseudoscience like to draw on the very human inclination to feel awed and taken aback in the face of the wonders of our universe. They use emotional language that plays on our feelings of ignorance and wonder and use this to convince us that we should listen to what they are saying. Science, they

Building your Skeptical Toolkit

say, doesn't know everything, and indeed they are right[*], so why not believe in the power of ancient healing crystals or the mystical energy of therapeutic touch?

Do the proponents also promote other fringe claims?

Does your acupuncturist also offer Reki sessions or ear candling? Does your Chiropractor sell homeopathic remedies? Has your Reflexologist ever commented on the state of your aura? It is not at all uncommon for promoters of one form of pseudoscience to also be supporters of other forms as well. If you are unsure if the modality you are being treated with is legitimate science then have a look at what else is on offer. If you come across something that you know to be bunk then there is good reason to be skeptical about the treatment you are getting. That said this may not always be the case. My local GP for example advertises an alternative medicine clinic on their appointment cards but this doesn't mean that the medical services they themselves offer are invalid, though it does annoy the hell out of me.

Does it employ logical fallacies in its defence?

Because many pseudoscientific and paranormal claims are often unsupported by any evidence and are themselves highly illogical their supporters are often forced to resort to the use of logical fallacies to defend them. We have already looked at a couple of logical fallacies in this list. The argument from antiquity, because something is old it must be right. The argument from popularity,

---

[*] To quote the brilliant comedian Dara Ó Briain "Of course science doesn't know everything. If science knew everything, it would stop."

Science! It Works, Bitches!

lots of people use it so it must work. The argument from common sense, everyone knows this, it's just common sense. Special pleading, the reason the results came back negative is because science can't test our claim. Those are just a few examples but there are lots more. If you are able to identify a logical fallacy in the arguments being put forward for a claim then it should raise a red flag. Now it is possible that the claim is accurate and that the person arguing in its favour is just doing a bad job of it, however if something is right then it should also be logical and there should be no need to use fallacies in its defence, even if that logic isn't always immediately apparent - I'm looking at you quantum physics.

Pseudoscience Bingo

Pseudoscience Bingo is a quick and easy way to identify whether you are dealing with real science or a whole load of bunkum. Simply evaluate the claim you are presented with and cross off any red flags that come to light. Should you score five in a row, be it horizontally, vertically or diagonally, then you have good reason to be skeptical, especially if your line passes through the centre square. Feel free to photocopy and distribute this page as needed.

It flies in the face of current knowledge or scientific thinking	Its supporters also promote other fringe claims	It is based upon an unknown form of energy	It was first announced through the press or popular media	It all sounds too good to be true
It relies upon corny marketing or celebrity promotion	It employs logical fallacies in its defence	Its supports claim it is being supressed by the authorities	It is not supported by good quality evidence	It is not self-correcting
It is based upon ancient knowledge	It claims to treat a wide range or all conditions	It is said to be based upon quantum physics	It has support that is political, ideological or cultural in nature	It claims to be "All Natural"
It makes appeals to ignorance or wonder	It fails to pass Occam's Razor or Hume's Maxim	It doesn't act like real science	It was created by a single person	It relies upon personal testimonies as evidence
It is not falsifiable or testable	It remakes appeals to history, public opinion or common sense	It uses meaningless terminology	It makes vague claims	It was created before the age of modern medicine/science

Building your Skeptical Toolkit

Science! It Works, Bitches!

Open mind or an empty head?
Do skeptics have an open mind?

To finish off this section of the book I want to flip to the other side of the coin and take a look at a few of the more common criticisms that you are likely to encounter once you declare yourself a skeptic and start using scientific thinking to back up your arguments. This is actually a good approach when looking at any issue. By looking at the negative things said about a subject as well as the positive points it allows you to see the topic under discussion from a different point of view. Also understanding the reasons why the complaints against a topic do or do not hold up under scrutiny will help to strengthen your understanding of the topic itself. It was this method of thinking that lead to my fascination with evolutionary biology. By studying the arguments both for and against this issue and, of course, the evidence supporting them, I was able to come to a much more rational conclusion than I would have had I just focused on one side or the other.

A lot of the negative reactions against scientific thinking and skepticism in general come from a misunderstanding of the things we have looked at in the previous chapters and there is one that you are likely to encounter more than any other, and that's the accusation that you have a closed mind. A true believer or proponent of some paranormal claim will often accuse you of having a closed mind because you won't believe the story they are telling you without good evidence. Likewise you may find yourself being told that you are not open minded simply because you demand more than just anecdotes before you will accept the latest pseudoscientific claim. So are they right? Do skeptics really lack an open mind?

Before we can decide if skeptics are open minded or not we need to understand exactly what it means to be open minded. Being open minded simply means you have a willingness to accept new ideas. Despite the fact that the accusation of closed mindedness is often aimed at scientists for their unwillingness to accept paranormal and pseudoscientific claims at face value, scientists are in fact often the most open minded of people. Science itself relies on the ability of scientists to accept new ideas. Not only that but some of the really big breakthroughs in science required not only a willingness to accept new ideas but completely new ways of thinking as well. If scientists were not willing to accept new ideas and instead clung to their existing beliefs then when new evidence arrived that contradicted those beliefs or showed them to be false it would either be overlooked or just dismissed out of hand. As we have seen, the ability to accept when you are wrong is a built in part of the scientific method.

Of course a willingness to accept every paranormal or pseudoscientific idea that so much as looks at you does not automatically make you open-minded. In fact this very willingness to uncritically accept everything that comes your way could actually lead you to being closed-minded. Let me explain.

Imagine that you and me were at my house along with two other people, one a hardened skeptic and the other a true believer in the paranormal. Suddenly we all hear a strange and spooky banging noise coming from the walls. The true believer immediately announces that this noise could only be caused by a ghost. The skeptic shakes his head and says that he doesn't believe this to be the case, that there is no reason to suspect ghostly activity. The true believer can't believe what he is hearing and accuses the skeptic of being closed minded and unwilling to

Building your Skeptical Toolkit

accept what is right before him. After all the evidence for a ghostly presence is right there for all four of us to hear.

So who is being closed minded here, the skeptic or the true believer?

You think about that for a moment, but while you do that I'll go downstairs and turn off the outside tap. Oh hang on a sec the banging noise has stopped? It looks like this unnerving phenomenon was actually caused by the plumbing in my old house rather than by any sort of supernatural agent after all. So how does that affect the answer to my question? Now that we know that there was in fact no ghost behind the banging noise what does that tell us about who was being closed minded?

Well, knowing the cause, it is fairly easy to see that, even though he was willing to believe in the supernatural solely on the basis of a strange banging noise, it is in fact the true believer who is being closed minded in this situation. He immediately jumped to a conclusion that fitted his pre-existing beliefs and was unwilling to consider an alternative explanation. This is the very definition of closed mindedness. As for the skeptic it is true that he stated that he didn't believe the noise was caused by a ghost, however that is not the same as dismissing the possibility out of hand. He was simply pointing out that a strange noise was not adequate enough evidence for a supernatural explanation. He may have sounded closed minded but in fact he did not rule out any possible cause for the noise, either natural or supernatural, while the true believer immediately ruled out the possibility of a natural cause.

Now you will have noticed that the skeptic in my fictitious example allowed for the possibility of the supernatural, even though he thought it unlikely, so what is wrong with believing in it? I mean isn't it better to believe in these things, to be "open minded" with regards to ESP, psycho-kinesis, alien abductions and faith healing? After all, it is possible that they are all true so what

169

is the harm in believing in them? Maybe our true believer has a point? Yes, it is true he was wrong in this instance but our skeptic can't prove that there aren't real ghosts out there so isn't it best to just believe in them? If it had turned out to be a ghost making the noise rather than just bad plumbing wouldn't we all be calling the skeptic the closed minded one right now? So maybe at the end of the day, and in spite of our example, it is the true believer who really is the open minded one?

So maybe the question we should be asking here is how "open" is the mind of someone who believes in the paranormal? Yes, it is true that it is possible that ESP might one day be found to be true. It is also possible that aliens really are visiting the planet and abducting people, after all most people are open to the idea of life elsewhere in the universe so why can't they have visited us? But here is the thing. Someone who is skeptical of these things can change their mind when presented with new and better evidence. If it were shown that someone really could repeatedly perform a psychic feat under test conditions then most skeptics would, maybe grudgingly, accept this. A true skeptic is always open to the possibility that they are wrong. On the flip side however, is there any amount of evidence that would convince a true believer that the paranormal really wasn't real and that they were wrong? Well if the past is anything to go on the answer is no. It is very rare indeed that a true believer will admit that they were wrong and that the paranormal doesn't actually exist. Most likely they will claim that more research is needed and, as Fox Mulder would often say, that the truth is out there. This inability to admit that they could be wrong once again shows that our true believer, and not our skeptic, is the one who is really closed minded.

So how can we best view new claims in a way that is truly open minded? Well despite what the Mad Hatter said our goal should

Building your Skeptical Toolkit

never be to believe "six impossible things before breakfast" and in order to stop this happening we need to apply a filter to the things we believe, and that filter is evidence. By running all new claims through this filter we do indeed run the risk of not believing some true things but we will also greatly reduce the number of untrue things we believe. Added to which a truly open mind is always open to new evidence and so the door is never really closed on a subject just because right now the evidence is not strong enough to convince us.

When we find ourselves faced with a new claim we should compare it against what we already know about the World. Does this new claim require us to completely change what we already understand about the World? Is the evidence being presented for it any good? Could we test this claim for ourselves or are we being asked to simply take it on faith? Remember, as we said before, the plural for anecdote is anecdotes and not evidence. If a hundred people reported that they were abducted by aliens then this may sound like convincing evidence that aliens are doing just that. However we should never just lump these claims together and count them as one big piece of evidence. We should judge all of these claims individually and see if they stand up on their own and without the support of the others. If individually they can all be explained by purely terrestrial means then we should not be swayed by the fact that collectively they sound impressive. Good evidence is what should convince us that something is true, not anecdotes, personal feelings or just the fact that it would be cool if it were true. Just because we want something to be true doesn't mean it is.

One last thing on this subject. It is important to realise that those on both sides of arguments regarding the paranormal and pseudoscience are always going to see the other side as closed minded and there is probably little you can do to change their minds, other than changing yours. The reason for this is that, if we are honest, we are all a little closed-minded. We all view the World around us through pre-existing biases, all of us tend to remember things that support our side of the argument and forget those that don't, and we all suffer from confirmation bias, which we will look at in more detail in a moment, to a lesser or greater extent. If we are honest, the only real difference between a skeptic and a true believer is how we filter the ideas that come our way. The reason that the skeptic is likely to be right more often than the true believer is that he generally has a better understanding of science and the scientific method and as such is better equipped to filter out the good evidence from the bad. As I said I would love magic to be real but unfortunately the evidence just isn't strong enough to support it, and the validity of the evidence is what counts. Brian Dunning summed this point up perfectly in his Skeptoid podcast episode on the subject.

> "So don't focus on buzzword labels like "closed minded" or "true believer". You can be both of those things and still be able to properly analyse evidence and draw a supported conclusion. You can also be guilty of neither fault, and yet be unable to distinguish a well-supported conclusion from mountains of poor evidence. Focus on the method behind the conclusion. Focus on the quality of evidence that supports the conclusion. The ad-hominem attack of "He's closed minded" says nothing at all about the quality of evidence."

That doesn't make sense
Science and common sense

Common sense is an important tool that we all possess, to a lesser or greater extent. It allows us to get through life in one piece, to make quick decisions and to size up the World around us. Common sense is generally viewed as knowledge that everyone knows and which allows us to make prudent choices and exercise sound judgement. It's common sense that tells us not to stick our hand in the fire or cross the road when the light is red. We all know not to go outside in a thunderstorm or to pet an angry dog. Why? Well it's just common sense.

The problem arises when this idea of common sense is taken too far and used as a reason to either dismiss the findings of science or to support paranormal or pseudoscientific claims. Science very often goes against common sense and this is seen as a reason not to accept what it has to say. Common sense tells you one thing while science tells you something else altogether. The common sense solution is easy to understand and it is something that everyone knows. Are you saying that everyone is wrong? Plus if it doesn't make sense then it can't be right, can it? Unsurprisingly conspiracy theorists and science deniers are very keen on arguments based upon common sense.

To better understand the disconnect that sometimes exists between common sense and science I want you all to raise your left hand high above your head. To those of you reading this book in public I apologise but I want you to lift your hand as well, no excuses. Ok now I want you to reach out with your hand and touch something. It doesn't matter what, the desk you might be sitting at, the top of your own head, this book, the person sitting next to you, though you should probably ask permission for that last one. Are you touching something? Good.

Ok so here's the kicker. You may think you are touching something and it may feel as though you are touching something but science tells us that you're really not. Not a single atom in your hand is touching a single atom in whatever it is you think you are touching. And it is not like they are just missing each other either. If it were possible to magnify those atoms up to a scale that we can better understand it becomes very clear just how far apart those atoms really are. Imagine that an atom in your hand was the size of a ping-pong ball. Now imagine that it is sitting in the middle of a football stadium. The closest atom of, for arguments sake, the desk your hand is resting on is located in the centre of another football stadium down the road. That's the kind of relative distance we are talking about, and in reality I am probably underselling it a bit. The only reason that your hand doesn't pass straight through the desk is because your hand, and everything else for that matter, generates an electric field. The electric field produced by your hand interacts with the electric field produced by the desk, and vice versa, and it is that interaction that creates the illusion that they are touching. Common sense tells us that you are touching something, and this is perfectly valid for everyday life, but science and reality tells us that in fact nothing is touching anything.

There are many scientific ideas that fly in the face of common sense but which the vast majority of us simply accept in our adult lives without a second thought. But think back to when these ideas where first presented to you as a child or when you yourself have tried to explain them to children of your own. Imagine you are holding a globe in your hand and you are explaining to a wide-eyed child that, despite everything their common sense tells

Building your Skeptical Toolkit

them, not only is the World not flat but there are people right this very moment on the other side of the planet who are completely upside-down compared to them. It is very likely that the child will not believe you at first and will, with the brilliance many loose as they grow, immediately point out the flaws in your argument. After all if there were really people living on the bottom of the Earth then surely they would fall off.

But while they are still trying to understand that bombshell let's go further. Imagine that now you tell them that not only is the Earth round with upside-down people living on the bottom but, even though it doesn't feel like it at all, it is also spinning at around a thousand miles per hour. On top of all that it is also hurtling through space around the Sun at a mind-blowing speed of 66,780 miles per hour. But we are still not done. Not only is the planet round, not only is it spinning, not only is it shooting around the sun but the whole solar system is rocketing around the galaxy at around 490,000 miles per hour. I wouldn't be surprised if the child thought you were making it all up, and yet it is all completely true.

Over the years common sense has been shown to be wrong many times. Common sense argues for a flat not a round Earth, a geocentric not a heliocentric solar system and a universe with us at the centre. It tells us that an aeroplane is too heavy to fly, that time is the same everywhere and that space is flat and constant. And when it comes to subjects like quantum mechanics common sense goes right out the window.

Evolutionary biologist and former Professor for Public Understanding of Science at Oxford University Richard Dawkins has put forward the argument that the reason these ideas seem so unnatural to us is because we evolved to live in what he calls a "middle-earth", and no it's not a Lord of the Rings reference. In our lives everything we deal with on a day-to-day basis falls into this middle ground. Things are neither galactically big nor atomically small. We travel at middle speeds, neither too fast nor too slow, and survive quite nicely at middling temperatures and radiation levels. We are simply not adapted to deal with the extremes the universe has to offer.

So does that mean that common sense is a bad thing? Well of course not. As I already mentioned, for the most part common sense is more than enough to get us through our lives without too much trouble. Being well adapted to a middle-earth existence really doesn't do us that much harm and it could be argued that it provides us with a privileged position from which to investigate the extremes that exist on either side of our everyday world. So what's the problem?

Well as we know common sense is comfortable and easy and people like things that are comfortable and easy. As such common sense is very easy to abuse and one of the easiest areas to see this is alternative medicine. Common sense tells us that things that make us feel bad and which we don't really understand are bad for us, while things that make sense and which make us feel happy must be good. As a result it is not at all uncommon for people to turn their backs on effective but invasive treatments such as chemotherapy and instead try some new age technique that panders to their self-image and tells them what they want to hear but which is otherwise ineffective.

Another example where common sense can do serious harm is with regards to the anti-vaccination movement. They often argue that sticking babies with needles simply can't be good for them, common sense tells us this. As a result more and more children are not being vaccinated and illnesses that were all but eradicated are making a comeback. Even worse, those children who can't be vaccinated for legitimate reasons are losing the protection provided by herd immunity due to too few people being

Building your Skeptical Toolkit

vaccinated to sustain the level of protection needed. Vaccines are arguably the biggest lifesaver ever developed by mankind and have successfully eradiated killers like smallpox from the face of the Earth, and yet arguments from common sense are undoing the work they have done to the point that we are currently seeing a resurgence in a number of diseases.

Common sense has its uses, I will not argue with that. However, common sense has been shown to be wrong many times and numerous things that many of us believe based upon common sense are just plain wrong. When presented with an argument based upon common sense you should never just accept it just because it sounds good and lots of people believe it.

After all common sense tells you that you are actually touching something right now. You can stop by the way.

Science! It Works, Bitches!

You can't prove I'm wrong
The burden of proof and asking the right questions

Another common reaction you will face from proponents, when you announce that you are skeptical of whatever claim they are making, is for them to tell you that it is up to you to prove them wrong and unless you can then their claim must be true. This is called shifting the burden of proof and is actually a logical fallacy, though one that I feel deserves special mention in relation to scientific thinking. Now as you will remember it is actually a misnomer to use the word proof in relation to science as nothing in science can ever be proven, only falsified. However for the sake of making it easier for me to explain things I will be using the word in the incorrect fashion in this chapter.

In science, as well as in law in most countries, the burden of proof always rests with the person making the positive claim. The reason for this is the fact that it is impossible to prove a negative.[*] Let's say I make the claim that reindeer cannot fly and that it is therefore impossible for them to be used to draw a sleigh across the sky, thus disproving the existence of Santa Claus as most people portray him. Now how would I go about testing this claim? Well the great James Randi proposed a thought experiment to do just that. He suggested we imagine that we take one thousand reindeer up to the top of a very tall building and start pushing

---

[*] Ok so I should have said practically impossible, as there are cases where it is possible to prove a negative. For example I could prove the negative that there is not an elephant in my pocket by looking in my pocket and comparing the relative sizes of said pocket and an elephant. But for the most part negative statements are far harder to prove.

Science! It Works, Bitches!

them off, one at a time, to see if any of them can actually fly. Now in our experiment all of the reindeer fail to fly and plummet to the ground. So does this then prove our case? Have we shown that it is impossible for reindeers to fly? Here's what Randi had to say:

> "What have we proven with this experiment? Have we proven that reindeer cannot fly? No, of course not. We have only shown that on this occasion, under these conditions of atmospheric pressure, temperature, radiation, at this position geographically, at this season, that these 1000 reindeer either could not or chose not to fly. (If the second is the case, then we certainly know something of the intelligence of the average reindeer.) However, we have not, and cannot, prove the negative that reindeer cannot fly, technically, rationally, and philosophically speaking."

In order for us to disprove the existence of flying reindeers we would need to test every single reindeer under every possible condition at every possible geographical location and at every possible moment in time. Clearly this is an impossible task. However to prove that reindeers can fly all we would need to do is produce one flying reindeer. Proving something is infinitely easier than disproving it.

The same argument applies to all paranormal and pseudoscientific claims. Repeated testing over many years has shown that acupuncture fails to work any better than what you would expect from a placebo, however this does not mean that we have proven that it does not work.[*] It is possible that under

---

[*] At this point one of my proof readers pointed out that acupuncture has been used as a way of reducing pain during child birth. For an informed opinion on this matter I turned to Professor of Complementary Medicine Edzard Ernst MD, who had this to say in an article on the subject, "There is some trial data supporting the use of acupuncture for reducing the types

Building your Skeptical Toolkit

some set of conditions and at specific places and times acupuncture can indeed perform miracles. Only by testing every single possible way in which acupuncture could work would we ever be able to prove that it does not. You also can't prove that no one can talk to the dead, that aliens haven't really abducted anyone or that Bigfoot isn't really out there somewhere. However, with definitive evidence, such as the corpse of a hitherto undiscovered hominid, you could go a long way to proving that these things do exist.

In all of the above examples we have a skeptic making a specific negative claim, but what if they are not making a claim at all? What if they are simply stating that they do not believe a claim presented to them?

Imagine we have a proponent of psychokinesis trying to convince us that they can move objects with their minds. They present no evidence of this ability and so we simply state that we do not believe them. Upon whom should the burden of proof rest in this case? Even ignoring what we have just talked about it should be fairly obvious. In this example, the only claim being made by the skeptic is a statement of disbelief in the claim presented to them. No evidence is required to back up the skeptic's claim of disbelief, save for the skeptic's word on the matter. In this example, the skeptic has something known as the benefit of assumption, meaning he needs no evidence to support his claim. However, the proponent is making a claim, and an extraordinary claim at that, which does require evidence in order to support it. Once more the burden of proof can only lay with the claimant.

Generally a good skeptic, when presented with a claim for which they have no specific counter evidence or way of falsifying, should always take what is known as the null-hypothesis. A null-

---

of [labour] pain listed above. Yet this evidence is far from being uniformly positive and is therefore not convincing."

hypothesis can be thought of as a neutral starting point. You are neither siding with the proponent in agreeing that their claim is true nor are you asserting that something else is in fact true. You are simply saying that, given the current evidence, you do not believe the claim presented to you. For example if someone were to make the claim "aliens abduct people" the null-hypothesis would be "aliens do not abduct people." The null-hypothesis is making no claim as to what is actually happening to people who believe aliens have abducted them, it is just a statement of disbelief in the positive claim. As such the null-hypothesis can be thought of as the negative that you cannot prove, which once again puts the burden of proof on the claimant.

Now there is another area that I think relates, at least tangentially, to the burden of proof issue. It is not uncommon for the proponents of pseudoscience to shift the focus of the debate onto the skeptic by asking loaded questions. Evangelical preacher Ray Comfort, founder of The Way of the Master ministry, offers up many such examples of loaded questions over on his website Atheist Central. Ray will often try to redirect discussions about the validity of evolution theory, big bang cosmology or the evidence for God with questions such as "Who created the universe?" A question like this immediately pulls the skeptic away from the null-hypothesis and, if addressed directly, may force them into making a positive claim for which they have little evidence. It also drastically limits their options.

The use of the word "Who" implies that a distinct individual being was behind the formation of the universe and immediately rules out naturalistic alternatives or multiple causes. Likewise, the word "created" implies that the universe was the result of some purposeful action and some level of intelligence as all known things that we would class as being "created" are done so with intent. The question is set up in such a way as to provide the skeptic with little choice but to give the answer the proponent is seeking. The two main ways to address a loaded question are to answer it as though it were correctly worded or point out that the question does not make room for all possible options and offer an alternate question that you would be willing to answer.

In this case replacing the word "who" with "what" makes way for all possible agents, both natural and supernatural, intelligent and benign, whereas changing out "created" for "caused" allows for the whole spectrum from purposeful action to pure random chance. A correctly worded question rules nothing out. The question "what caused the universe" is therefore a much better one and covers both a creator God and pure naturalist means.

Science! It Works, Bitches!

Going beyond the realm of science
Science and the supernatural

This brings us nicely to the last common complaint against science that I want to address. That being that science is based on naturalism, the unproven assumption that nature is all there is and that it automatically rules out the possibility of a supernatural cause for anything and as such would never find one even if it existed. Let's look at the various parts of this complaint in turn.

Firstly this complaint is partially correct; science is indeed based upon naturalism, but importantly upon a very specific form of naturalism. The kind of naturalism employed by science is called "methodological naturalism". Scientists that use methodological naturalism limit the scope of their research to the study of natural causes not, and this is the important bit, because they completely rule out the possibility of a supernatural cause but because the inclusion of supernatural agents or forces in scientific research has consistently failed to produce reliable or verifiable results. This form of naturalism can be thought of as a tool that scientists use and does not make any claims as to the truth of supernatural causes. In this respect it differs from philosophical naturalism, which does takes the point of view that the supernatural does not exist and that the natural is all that there is. Critics of science often confuse these two different ways of thinking, though in their defence there are some proponents of science who do subscribe to both of these forms of naturalism.

So why does science use methodological naturalism in the first place? Why not include the supernatural when looking for the cause of things? Well the reason that science only focuses on

Science! It Works, Bitches!

natural causes and appears to ignore the supernatural is because the supernatural is, definitionally, beyond our ability to detect and as such not something that can be reliably quantified or measured. The dictionary defines the word in this fashion:

> Supernatural: of, pertaining to, or being above or beyond what is natural; unexplainable by natural law or phenomena; abnormal.

There are many scientists out there who do believe that there is more to the universe than just the natural, take all the scientists who also have religious beliefs for example. However, this criticism against science keeps coming up and so we have to ask if having a naturalistic view of the World is really a bad thing as the critics claim? Well we can say quiet definitively that no, of course it's not. All of us, including those who speak out against it, use a form of naturalism every day. If people assumed that everything that happened to them did not have a natural explanation then they simply would not be able to cope in the real world. Why eat when you believe that supernatural forces rather than food keep your body alive? Why bother looking both ways before crossing the road if you don't believe in the naturalistic physical forces that would act upon you should a car hit you? More importantly methodological naturalism works. This way of thinking has given us modern medicines that have eradicated global killers like smallpox from the planet. It has given us clean drinking water, which itself has saved countless lives, and has allowed us to grow more food than would otherwise be possible. It has given us technologies ranging from the printing press to the internet and has even put a man on the moon, several times. Supernaturalism on the other hand does not work. Science, as we have already discovered, is based upon evidence. If two scientists disagree about something then they can point to the evidence as a way to resolve the debate. You can't do that with supernatural things. Let's imagine for a moment that we did include the supernatural in science. How do you know for sure if a supernatural cause is influencing an observed phenomenon or not, and if it is how do you distinguish one possible supernatural cause from another? Maybe gravity isn't caused by a bending of space/time but in fact is due to fairies pulling objects closer together. Perhaps electricity is actually the result of Zeus throwing lightning bolts and yes maybe all life on Earth was just spoken into existence. But how can you demonstrate any of these things one way or another when they exist by their very, if you will excuse the use of the word, nature in a state that is completely undetectable in any way, shape or form? Is it possible that the supernatural exists? Yes, but for it to have any effect within our Universe whatsoever it would need to interact with it in a naturalistic way and as such be detectable by science. The science website Talk Origins also addresses this question regarding methodological naturalism and deals with the issue of the intrinsic uncertainty of any supernatural claim in a more direct fashion than I have done and ends its article with these words:

> "People tend to have different and incompatible ideas of what form supernatural influences take, and all too often the only effective way they have found for reaching a consensus is by killing each other."

The second part of this criticism is that because of the naturalistic approach taken by science it will miss any supernatural explanations that do exist. So the first question to ask is what exactly do people mean when they talk about the supernatural?

Science! It Works, Bitches!

As we saw earlier, the supernatural is, by definition, something that is unexplainable and beyond what is natural. But how can something that is beyond natural have an effect on the natural world. Something that has an effect on the natural world has to be part of nature itself. As such anything we see having an effect on the natural world is considered natural by its very definition. If the supernatural does affect our natural world it must do so in natural ways. So not only is a possible supernatural cause something that is impossible to usefully define but it would also be impossible to observe or detect in any way. As such, the only way to conclude that a supernatural force is at work is to first eliminate all natural causes for the event in question. But this results in us trying to prove a negative, that there is not a natural cause for effect X. We can never say for sure that we have eliminated all possible natural causes, as there could always be a natural cause that we are not aware of or which is currently beyond our ability to detect with our current level of technology. So even if a supernatural explanation is the correct one science could never actually reach the point of confirming it.

But let's say, for the sake of argument, that we do come to the conclusion that event X is caused by something supernatural. It is completely impossible for us then to determine which of the plethora of possible supernatural causes is responsible for the event as they are all equally beyond our ability to test or even detect. Supernatural things cannot ever be demonstrated to exist by science and as such must be taken on faith. Let's take thunder and lightning for example. Before we had an understanding of the natural causes of this meteorological phenomena many different religions over the centuries attributed it to the actions of numerous different supernatural beings. The Greeks believed that Zeus was responsible while the Roman's attributed it to Jupiter. The Norse were sure it was the work of Thor and the Celts argued that it was down to Taranis. In fact a very quick search on Wikipedia will reveal a list of 50 different gods that were believed to be responsible for thunder and lightning. How are we to distinguish which, if any, of these deities is responsible? The simple answer is we can't, and yet many people believe that their faith in a supernatural cause should be seen as equally valid, or often in their minds more valid, than natural, empirical, testable scientific evidence.

But ok, again let's say that they are right and our naturalistic approach to science does cause us to miss out on the actual supernatural cause for something, what then? Well, to be blunt, would it really matter? Even if something does have a supernatural cause we will never ever be able to measure or test that cause as it is completely beyond our ability to do so, and not just presently but by the very definition of the term this will always be so. We can't use our knowledge of this supernatural cause to see if it has an effect on other natural things and we can never ever hope to utilise this supernatural force to improve our lives as we only have natural things with which to work. In short, a supernatural cause would, to all intents and purposes, be completely and utterly useless to us in all possible ways.

As I said above this doesn't mean that supernatural explanations are not evaluated to the extent that it is possible to do so by science. Science has been used to evaluate supernatural claims regarding faith healing and various faith based alternative medicines for years. It can also be used to test specific faith based religious claims such as the global flood and the age of the Earth as these things, though they are said to have a supernatural cause, are claimed to have caused naturalistic effects and as such the evidence, or lack thereof, will be there for science to examine. Let me give you a practical example that I feel shows very clearly why the use of methodological naturalism by science is the

Science! It Works, Bitches!

correct approach and why the supernatural, even if it does exist, is of little help with regards to our understanding of the universe. To do this I am going to use something called the Fallacy Model, which was created by Tracie Harris, the co-host of the public access TV show The Atheist Experience. As you might guess from the title of the show Tracie used this model as a way of demonstrating specifically why attributing anything to the power of God or Gods doesn't get us anywhere, however it works just as well with regards to any supernatural agent and it is in that capacity that I will use it here.

Fig 1. Existent Dice, Non-Existent Dice and Transcendent Supernatural Dice

Imagine that you have three jars laid out in a line in front of you. In the first jar you have Existent Dice, that being dice that really exist. In the second jar you have Non-Existent Dice, or to put it another way you have an empty jar. And finally, in the third jar you have Transcendent Supernatural Dice, or dice whose properties put them outside of the confines of the natural universe. Now for the sake of argument we are going to say that the dice in the third jar, the Transcendent Supernatural Dice, do in

Building your Skeptical Toolkit

fact exist, that the jar really does contain dice that are transcendent and supernatural in nature. With me so far? Good.

Let's start by looking at the first jar, the one with the Existent Dice. We can say that the dice in the first jar exist for a number of reasons. We can see them for a start and if we were to give the jar a shake we would be able to hear the sound they make when they hit against the side of the jar. We could weigh the jar and then take the dice out and weigh it again to see if the dice have mass. We can also feel the dice, detect that they have a texture and are solid to the touch. If we had the ability we could run experiments on the dice to find out what they were made of, what their chemical and molecular structure is etc. In short there are numerous things we can do to verify that the Existent Dice do in fact exist.

Now imagine that you close your eyes and that someone rearranges the jars so that they are still in a line but in a different order to how they were before. When you open your eyes again you are asked to divide the jars up into two groups. One group will be for jars that contain things that exist and the other group will be for jars that contain things that don't exist. Remember we have agreed to say that the Transcendent Supernatural Dice do actually exist and so as such should go into the group of jars that contain things that exist. So the question is how do we do this? How do we divide these three jars and their content up into the two groups, one for jars that contain things that exist and one for jars that contain things that don't exist? Feel free to think about it for a bit before you read on.

Now obviously we can start by sticking the jar containing the Existent Dice straight into the group of jars that contain things that exist, for the various reasons we mentioned above. We can see the dice, hear them, touch them etc. But what about the other two jars? How do we differentiate the Non-Existent Dice from the Transcendent Supernatural Dice?

The answer is that we can't. The Transcendent Supernatural Dice have the exact same properties as the Non-Existent Dice. There is no discernible difference between the two. The supernatural is, as we covered above, that which is beyond our ability to detect. As such for the Transcendent Supernatural Dice to have any properties that we could measure in order to compare them to the Non-Existent Dice they would no longer be classed as supernatural. The supernatural, at least as we are using the term in this example, has the same properties as the non-existent. The problem is that we have already stated that the Transcendent Supernatural Dice do in fact exist, and so that raises a very important question. What exactly do we mean by the word "exists"?

By stating that the Transcendent Supernatural Dice do in fact exist even though they have the exact same properties as the Non-Existent Dice we have, essentially, eliminated any meaning from the term "exists". We can no longer differentiate the existent from the non-existent if something with all the exact same properties as something that is non-existent does, in fact, exist. Still with me?

As such if I say that magical invisible pink unicorns exist and are right this moment controlling your mind how would you prove me wrong? If existence isn't reliant on any manifesting traits (such as visibility, weight etc) then I can simply state than something exists and that it simply doesn't manifest, at which point again the term "existent" becomes completely meaningless.

Let's put to one side for the moment the jars with the Transcendent Supernatural Dice and the Non-Existent Dice and focus once again on just the one with the Existent Dice. Now if I was to ask you to guess how many dice there were in the jar you

Building your Skeptical Toolkit

would probably say, judging from the picture on the previous page, something along the lines of thirty or so. Take a moment to have a look at the picture and come up with whatever number you feel comfortable with. Let's say I also make a guess and I say that there are 257,000 dice in the jar. Now I don't know any more than you do how many dice there actually are in the jar but does that mean that my guess of 257,000 dice is just as valid as your guess of around thirty odd?

Well of course the answer is no, but why not? Well it's because of what we do know. We may not know how many dice there are in the jar but we do know, and admittedly this would be easier with an actual jar of dice rather than a picture, that the dice appear to be the same size and that they take up around half to two-thirds of the space inside the jar. Given this information we can say, with an incredibly high degree of certainty, that it would be impossible for a jar of this size to hold 257,000 dice of the size all these dice appear to be. Just because we don't know for sure one piece of information, in this case how many dice there are in the jar, it doesn't mean that any and all guesses as to the number are equally valid.

Let's now swap jars and take a look at the jar that contains the Transcendent Supernatural Dice instead and once again I'll ask the same question. How many dice do you think are in the jar? For the sake of argument let's say that once again you guess a number around the thirty mark and that once again I guess 257,000. As before neither of us actually knows for sure how many dice there are in the jar and so we can again ask if that means that both of our guesses are equally valid? Well this time the answer to that question is yes. The reason the answer is yes is because all the information needed to judge the validity of the two guesses is no longer there. We don't know, and in fact have no way of knowing, the size of the Transcendent Supernatural

193

Science! It Works, Bitches!

Dice in the jar or how much of the space in the jar they take up. Because of this any guess is just as valid as any other guess.

Let's bring this back to the real world. If someone makes the claim that a supernatural cause for something is just as likely as a natural cause for something then upon what information are they basing that claim? For example if a Young Earth creationist makes the claim that their God created the World in six day just 6,000 years ago but with the appearance of age and I say that, actually, it was created last Tuesday by those pesky magical invisible pink unicorns who also created all our memories so that we would think it was older, then upon what information can you judge which of these claims is the correct one? Why is one more likely than the other? Is one more likely than the other or are both equally possible given what information you do know about the possible supernatural cause?

Alternatively, if I say that the current state of the universe was the result of the Big Bang around 13.5 billion years ago I may still be wrong, just as a guess of thirty Existent Dice in the jar may be wrong, but at least I am able to point to the evidence and information that supports my claim which you could then examine for yourself. We understand things about the universe, such as gravity and matter and the laws of physics, and as such a claim about the universe based upon this evidence is not the same as a claim about the universe based upon the supernatural. Now the obvious follow on question from this is what caused the Big Bang? If you asked me this I would have to tell you that I don't know, but ask someone else and they might state that the cause was those magical invisible pink unicorns. Is it possible that magical invisible pink unicorns were the cause of the Big Bang? If we are honest we have to admit, as we covered earlier, that yes it is possible. However, if we admit that we have to also admit that any other supernatural claim that we can think of is equally

Building your Skeptical Toolkit

possible as they all have the exact same amount of evidence in their favour, which is to say no evidence what-so-ever. If there is no evidence for what came before the Big Bang and in fact no current way even to get this evidence then the honest answer to that kind of question is to admit that you don't know. Any other answer is nothing more than a guess and is indistinguishable in validity from any other guess. Just because no one knows what happened before the Big Bang it doesn't make any guess more of a possibility than any other guess, no matter how detailed or involved that guess may be. Remember it is always about the evidence.

And if we think about it for a moment the answer of "I don't know" may actually be more accurate than any other answer you can give to this kind of question. When you say that you don't know you are not assigning any properties or characteristics to your answers and as such there are no properties or characteristics that can later be shown to be wrong, as you are simply stating the null hypothesis. If you say that it was caused by magical invisible pink unicorns or by some god or gods however then you are assigning properties or characteristics to your answer that in reality you cannot possibly know and which could, should the real answer ever become available, be shown to be wrong. So in the end the answer of "I don't know" is not only the most honest but it also can be the most accurate.

And that's where I want to end the first part of this book. Hopefully I have managed to get across the basics of scientific thinking and why it is both vitally important and inexorably linked to being a good skeptic. Also by looking at some of the common arguments against this method of acquiring knowledge I hope I have made it clear why science offers us a far better way of finding truth than any other route yet conceived by mankind. The philosopher A C Grayling put it much more bluntly and far more

Science! It Works, Bitches!

succinctly than I have when he wrote the following in his column in The Guardian.

> "On one side are those who inquire, examine, experiment, research, propose ideas and subject them to scrutiny, change their minds when shown to be wrong and live with uncertainty while placing reliance on the collective, self-critical, responsible and rigorous use of reason and observation to further the quest for knowledge.
>
> On the other side are those who espouse a belief system or ideology which pre-packages all the answers, who have faith in it, who trust the authorities, priests and prophets, and who either think that the hows and whys of the universe are explained to satisfaction by their faith, or smugly embrace ignorance. Note that although the historical majority of these latter are the epigones of one or another religion, they also include the followers of such ideologies as Marxism and Stalinism – which are also all-embracing monolithic ownerships of the Great Truth to which everyone must sign up on pain of punishment, and on whose behalf their zealots are prepared to kill and die."

But I don't want to end this part of the book on that slightly negative note and so I will once more turn to the great wisdom of the late Carl Sagan whose elegant summary of the arguments presented thus far borders on the poetic.

> "There is no other species on Earth that does science. It is, so far, entirely a human invention, evolved by natural selection in the cerebral cortex for one simple reason: it works. It is not perfect. It can be misused. It is only a tool. But it is by far

the best tool we have, self-correcting, ongoing, applicable to everything."

As I said on the title page of this section, Science, it works bitches.

Science! It Works, Bitches!

Part Two

Your Brain Hates You!

A few ways in which your brain will fool you if you let it

Building your Skeptical Toolkit

Tell me something I don't already know
The power of confirmation bias

A strong case could be made for the idea that your brain hates you. I know, it seems impossible, after all you spend so much time together and you no doubt believe you get on fairly well. However, as you will learn over the next few chapters, your brain is more than happy to fool you if you let it. By teaching you about some of the more common tricks it likes to play on you I aim to equip you with the tools you will need to spot these deceptions and, by recognising them when they occur, counter their effects.

Once you have learnt how to identify these brain tricks in yourself you will also be able to spot when other people have fallen foul of them, either in the arguments they make or the data and "evidence" they present to you for a claim about which you are, probably quite rightly, skeptical. But be aware, this is an on-going process and even knowing that these tricks exist won't stop your brain from trying to get away with them. After all the price of good critical thinking skills is eternal vigilance.

The first brain trick we're going to look at is one that everyone falls for at some point and one that forms a foundation for many of the other ways your brain likes to fool you. We're going to start by looking at confirmation bias, but then I expect you already knew that.

At first glance confirmation bias appears fairly simple and straightforward. However the more you learn about it the more you start to realise what a strong effect it has on a great many aspects of our daily lives. Psychologists generally use the term "confirmation bias" to describe the way that people tend to

Your Brain Hates You!

selectively collect new evidence in order to support their pre-existing beliefs and attitudes. However it can also be used to describe the biased interpretation of evidence as well as the selective recall of memories.

Let's start by taking a look at these four cards:

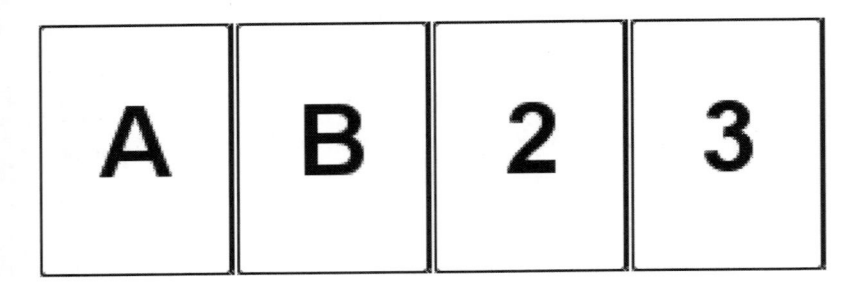

| A | B | 2 | 3 |

Now as with the jars and dice in the previous chapter, this would be easier if these were actual cards rather than just pictures, but I want you to image that they are real cards and as such have two sides. On one side is a letter and on the other is a number. Now I'm going to presented you with the hypothesis that all cards with a vowel (A,E,I,O,U) on one side have an even number on the other side. In order to test this idea you can turn over only two of the cards. Think about it for a moment and decide which of the two cards you would turn over, maybe even write it down if you have pen and paper to hand so that you have a note of what you decided for when we come back to this in a little while.

As I only briefly explained it above so let's start by looking at exactly what confirmation bias is. Confirmation bias is the very human tendency to search out or look at information, generally new information but not exclusively, in a way that confirms things that you already believe and to ignore or downplay information that contradicts those beliefs. One of the most common examples of this that many of you will have come across has to do with hospitals and the full moon. Ask almost any doctor and they will

Building your Skeptical Toolkit

tell you that when there is a full moon things get crazy. They all remember times when there was a full moon and things were extra busy but they tend to forget those times when the moon was full and things were quiet or when things were manic and the moon was far from full. This is a great example of confirmation bias.

Another example is found in the methods of so called psychics. At a psychic reading a psychic may ask three hundred leading questions, make fifty vague statements of which maybe ten or so are accurate and yet you will leave with the impression that the psychic knew everything about you. This too is due to confirmation bias. You remember the hits and either ignore or downplay the misses. Psychics rely on this very natural human trait for their business. But I suspect most of you have never been to a psychic, well horoscopes work in exactly the same way. They throw out lots of vague statements some of which will eventually sound fairly accurate, mainly because they apply to most people. People tend to remember the times when their horoscopes seemed accurate and forget the times when they were off by miles.

In 1960 cognitive psychologist Peter Cathcart Wason developed one of the first tests to examine confirmation bias that he called the 2-4-6 problem. The way this works is very simple but also very clever. The subjects of the experiment were given a series of three numbers, known as a triple, and told that this triple conformed to a certain rule. The aim of the experiment was for the subject to work out what this rule is. They would start with something like this:

203

Your Brain Hates You!

Sequence	Fits the instructor's Rule?	Guess the instructor's Rule	How Sure?
2,4,6	Yes	Counting up by two's	50%

In order to work out what the rule was they had to generate their own triple and would be told by the person running the experiment if their triple fitted with the rule or not, after which they were asked to state how confident they were that they had guessed the rule correctly. They were asked to repeat this process until they were sure they had the correct answer at which point they were told to announce the rule they had come up with. Most people taking part produced something like this:

Sequence	Fits the instructor's Rule?	Guess the instructor's Rule	How Sure?
2,4,6	Yes	Counting up by two's	50%
6,8,10	Yes	Counting up by two's	60%
20,22,24	Yes	Counting up by two's	70%
3,5,7	Yes	Counting up by two's	80%
25,27,29	Yes	Counting up by two's	90%
200,202,204	Yes	Counting up by two's	100%

Now all of these triples seem to verify that the rule is in fact "counting up by two's", but that is only because all of the suggested triples are testing for a positive result and thus are just confirming what the subject already believes the rule to be. In order to get to the actual answer they would need to come up with triples that do not fit with what they believe the rule to be. For example they might try things like this:

Sequence	Fits the instructor's Rule?	Guess the instructor's Rule	How Sure?
2,4,6	Yes	Counting up by two's	50%
10,20,30	Yes	Counting up by multiples	60%
100,500,894	Yes	Counting up with all even numbers	70%
1,9,20	Yes	Counting up	80%
27,13,4	No	Counting up	90%
55,2,999	No	Counting up	100%

What the experimenters found was that while the actual rule was "any ascending sequence" the subjects would generally come up with much more complex rules due to the fact that they tended to test only "positive" examples, in other words triples the subjects believed would conform to the rule they had come up with and which confirmed their hypothesis.

You will probably remember from earlier in the book that we looked at the idea that nothing in science can ever be proven, only falsified. In science it is generally accepted that the best way to demonstrate that a hypothesis is accurate is to attempt to falsify it. This approach of testing for a negative result is in part designed to eliminate the chance that your results are due to confirmation bias. As the above experiment shows, by only testing for positive results it is completely possible to get a result that is entirely consistent with your hypothesis and yet at the same time is totally deceptive and gets you no nearer to the truth.

Ok so now that you have a better understand of what confirmation bias is let's go back to those cards from the start of the chapter. I expect all of you to get the right answer now, which is why I asked you to make of the note of the answer you came up with before I explained how confirmation bias works. No cheating now.

The hypothesis I presented to you was that all cards that have a vowel on one side have an even number on the other. In order to test this I asked you to pick two cards to turn over to see if the hypothesis is correct or not. Now once again this is an actual experiment, also created by Wason, that has been conducted many times over the years and the results have shown that the vast majority of people pick the A and 2 cards. As such let's start by turning over these two cards.

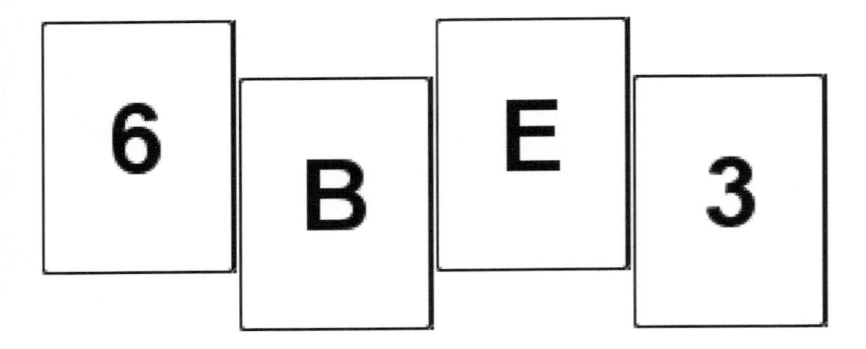

6   B   E   3

So there we go, we have proven our hypothesis. Cards with a vowel on one side have an even number on the other. But wait; is that what we have done? Let's turn over another card, this time the 3 card, and see what we get.

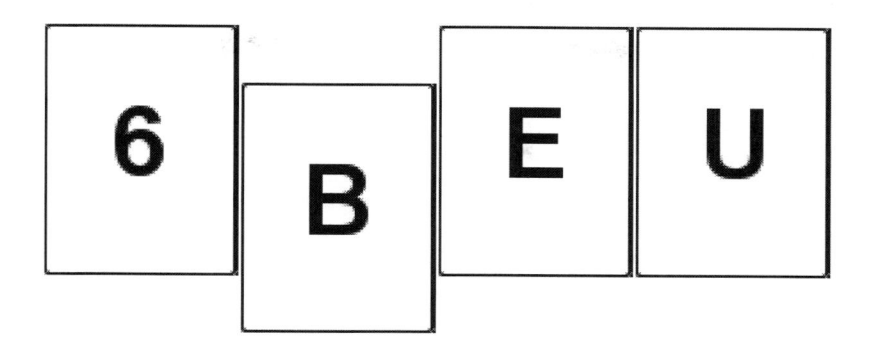

```
┌─────┐ ┌─────┐ ┌─────┐ ┌─────┐
│     │ │     │ │     │ │     │
│  6  │ │     │ │  E  │ │  U  │
│     │ │  B  │ │     │ │     │
│     │ │     │ │     │ │     │
└─────┘ └─────┘ └─────┘ └─────┘
```

So the 3 card also has a vowel on the other side and thus our hypothesis has been shown to be false. So which cards did you pick? If like most people who do this experiment, myself included, you picked the A and 2 cards then you have fallen foul to confirmation bias. You chose to turn over two cards that would confirm the hypothesis rather than choosing one that would potentially falsify it. This is something that many people do in everyday life. People have a tendency to seek out the things that will confirm what they already believe to be true and shy away from those things that would prove them wrong. For example, as mentioned above, someone who believes in psychics will be more likely to point to the hits and ignore the misses, thus confirming their pre-existing bias, than someone who looks at the situation with a truly skeptical mind.

Before we move on to look at some of the ways in which confirmation bias can affect the way in which we think and the opinions we hold let's take a very quick look at another set of cards. Imagine that we are playing a game of poker with a couple of friends. The cards are shuffled and the hands dealt. In the first two rounds you are dealt the following two hands:

Within the confines of our poker games I am sure you would be much happier about being dealt that second hand than the first, however, what I want you to think about is which of the two hands has the greatest odds of being dealt? Think about it for a moment before reading on.

Ok I won't give you too long to think about it and I will be honest and say that I doubt any of you fell for my little trick. By wording the question in the way that I did I implied that the answer was one hand or the other when in fact, as I am sure you guessed, the answer is that the odds of getting either hand are exactly the same. Again this is another, slightly different, form of confirmation bias. We assign more meaning to the second hand, the royal flush, and so many people assume that the odds of getting this hand are higher than those of being dealt the first hand. In fact the odds of getting any five card hand in a game of poker are identical. It is merely the meaning that we assign to the hands that differs. Coming back to our psychic they might say something like "I'm getting someone who died of something in the chest or stomach area." Now you know that your Grandfather died of a heart attack and so you assign this as the meaning to what was in fact a very vague and generalised statement from the psychic. The psychic then "confirms" that they were talking about your Grandfather dying of a heart attack all along when in fact it was you who assigned the meaning not them.

Confirmation bias is also the force behind a number of other things we may come across in day-to-day life. For example maybe you know two people who have strong but opposing views on a particular issue or topic and who seem to see every new piece of information that comes along as support for their side of the argument. As such overtime their beliefs become more and more disparate. This is an element of confirmation bias known as attitude polarization and research seems to show that it has a stronger effect on people who have previously stated their beliefs out loud for others to hear.

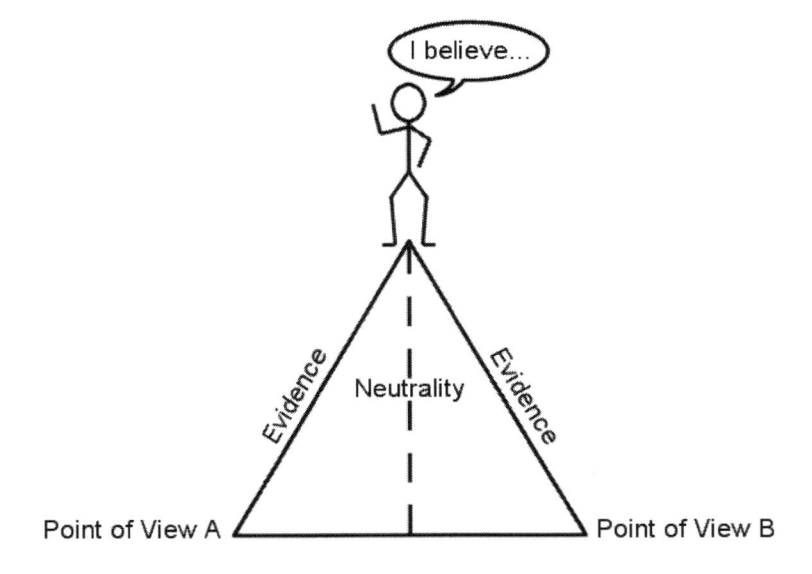

The diagram above shows this in a very clear and simple way. With any belief you always start from a neutral point of view.

Your Brain Hates You!

Stating an opinion on the issue publicly puts you squarely on one side of the arguments neutrality line or the other. The more you state your support for the belief and the more evidence you find in favour of it the further you move away from neutrality and those who hold the opposing belief. As such finding common ground becomes more and more difficult.

Similarly there are those people who seem to cling to their beliefs no matter how much evidence comes along that shows them they are wrong, and again the effect appears stronger with those who have publicly made their beliefs known. Experiments have been conducted in which people are shown faked evidence for something, for example a criminal's involvement in a crime, and are then asked to state their opinion as to the guilt or innocence of that person. They are then informed that the evidence they were shown was faked. Interestingly, and somewhat worryingly, the results of such experiments show that people tend to hold on to the opinions they formed when they first saw the faked evidence even after knowing that it is fake.

On top of this is appears that the order in which evidence is introduced can have an effect upon how much weight we assign to it. For some reason we tend to place more value on the evidence that is presented to us first than we do on the evidence we are shown later on, even if all the evidence is of equal importance. A simple example of this can be demonstrated with a list of words. Let's say you are asked to give your opinion of someone based upon the following description:

Intelligent, Industrious, Impulsive, Critical, Stubborn, Envious

The evidence shows that people will form a much more positive opinion of a person described in this way than if they were described as:

Envious, Stubborn, Critical, Impulsive, Industrious, Intelligent

Even though the words in both descriptions are identical the order in which they are presented seems to play a big part in how they come across. Bear that in mind next time you are writing your CV.

Now these issues may seem trivial but imagine the effect they can have when played out somewhere where the truth of the matter is of the highest importance, such as a court of law. Evidence from mock trials shows that jurors on the whole tend to be more likely to convict someone of a crime based upon a false confession even if all the other evidence shows that the person is innocent and, more importantly, even if the confession is shown to have been false later in the trial.

That is the power of confirmation bias and as I said at the start of the chapter it is something that affects us all from time to time. As such it would be remiss of me to end this chapter without mentioning the recognised cognitive bias known as bias blinding or the bias blind spot. As the name implies this relates to the fact that most of us, while often able to spot bias in others with ease, find it difficult to recognise the same bias in ourselves.

An experiment was carried out at Princeton University by psychologist Emily Pronin and her colleges on what is often referred to as the better-than-average effect or illusory superiority. In the experiment people were asked to rate themselves with regards to various qualities and it was found that most people overestimate their positive qualities and abilities and underestimate their negative qualities. Specifically when asked to rate themselves with regards to how biased they were most people rated themselves as being far less affected by the biases presented to them than the average person.

Your Brain Hates You!

Stop for a moment and consider your own thoughts on this question. How do you rate yourself with regards to bias? Do you consider yourself more, less or equally biased compared to the average person? I'm willing to bet that you consider yourself below average in this regard. I know I do.

Pronin suggested that the reason for this is found in the different ways that we assess bias in others and in ourselves. With other people we tend to use overt behaviour, that's obvious or easy to recognise behaviour, to draw our conclusions as to the persons standing in regard to a given bias. However when we look at ourselves we become introspective, turning our thoughts inward and searching them and our feelings for signs of bias. The problem is that biases generally operate unconsciously and as such are unlikely to be revealed by such introspection. As a result people mistakenly assume that therefore they are less susceptible to biased thinking that other people, ironically falling victim to biased thinking in the process.

As a good skeptic your aim is to mitigate the effects of confirmation bias in your own life and keep your eyes open for how it plays out in the lives of others. To help you with this here is a simple, three step approach to reducing confirmation bias in your life:

1. Try to judge every new piece of information that comes your way on its own merits as opposed to how it fits with your pre-existing beliefs.
2. Be open to other points of view even if your instinct is to simply dismiss them out of hand.
3. As identifying biased thinking in ourselves is something we all find difficult simply by pass the process by asking other people to let you know if you are responding in a biased fashion. Find a friend that you trust and ask them

for their honest opinion about your thinking on a given issue. As I mentioned above we generally find it easy to identify bias in others so make use of this fact to improve your own thinking.

If you can identify the ways your thinking is affected by your personal biases then you can take this into account when evaluating any new claim presented to you. Do this and it will help you obtain that one bias skeptics should strive for, a bias towards the truth.

Your Brain Hates You!

I know it's wrong, but...
Cognitive Dissonance

So we have looked at how people can, and do, favour information that fits with their pre-existing beliefs and how they can even hold onto those beliefs once the foundation for them has been shown to be false. Now let's take it a step further. Not only do people regularly continue to believe things that they know to be false but it is not at all unusual for people to believe two completely contradictory things at the same time. This, as you would imagine, can cause some mental discomfort that can result in feelings of anger, guilt, shame and stress, as well as other negative emotional states. Psychologists call this discomfort Cognitive Dissonance and, as the theory goes, believe that people are driven to try and resolve this mashing of conflicting ideas, or "cognitions", in one way or another. These feelings of discomfort are strongest when dealing with issues that relate to our beliefs about ourselves. For example you no doubt believe you are a good person and yet from time to time you may do something bad. This causes a conflict between the belief you have about the type of person you are and the action you carried out and the discomfort this causes is cognitive dissonance.

Now cognitive dissonance is not a bad thing in and of itself. In fact, it is a useful way of identifying flaws within your thinking or, as in the example above, may even make you a better person. If something is causing you mental discomfort then there is probably a good reason for it and it is worth your while to try and resolve the conflict. The problem with cognitive dissonance however arises in the way that people go about this resolution.

Your Brain Hates You!

Psychologists recognise three distinct ways of resolving instances of cognitive dissonance. These are:

1. Changing our behaviour or beliefs.
2. Forgetting or reducing the importance of one or both of the conflicting ideas.
3. Justifying our behaviour by introducing a new idea to the mix.

The first two of these provide constructive ways of dealing with your conflicting beliefs, to a lesser or greater degree, while the third, though on rare occasions useful, tends to lead people to irrational conclusions or actions. It is this third approach to the problem of cognitive dissonance that you are most likely to encounter on your skeptical journey, but let's address them in turn shall we. It is also worth noting that these approaches are not mutually exclusive and that people may use a combination of them to resolve a case of cognitive dissonance.

First we need a situation that will result in a feeling of cognitive dissonance. One of the most popular examples, and for good reason I feel, involves a person who smokes. This person is intelligent and well informed about the dangers of smoking and yet smokes anyway. Now, ignoring for the time being the addictive nature of smoking, the two conflicting cognitions here are as follows:

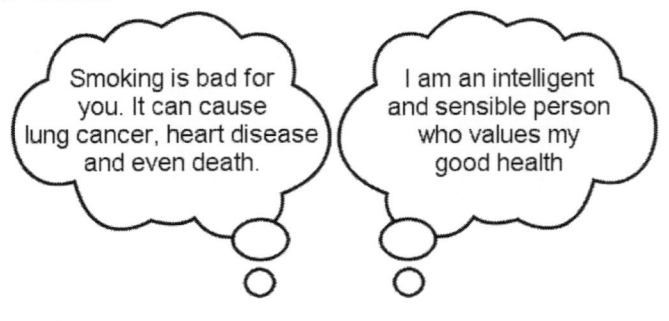

> Smoking is bad for you. It can cause lung cancer, heart disease and even death.

> I am an intelligent and sensible person who values my good health

Building your Skeptical Toolkit

Now clearly these are conflicting beliefs and are likely to cause any smoker some level of cognitive dissonance. If you truly are an intelligent and sensible person who cares about your health then why would you be doing something that you know puts your health at risk? As stated there are three ways of dealing with this problem.

Changing our behaviour or beliefs

In the example of smokers this presents you with one of the most common and most logical approaches to the problem. You just stop smoking. You know it is bad for you and you are a sensible person who cares about their health. The logical thing to do is to stop doing the thing that puts your health at risk. You continue to hold both of the beliefs you started with, only now they are no long a source of conflict. Cognitive dissonance gone.

Alternatively you could stop believing one of the conflicting beliefs. Now in this example this may seem a slightly irrational thing to do as neither of the beliefs here are fundamentally flawed. Smoking does cause health problems, it says so right on the box, and there is no reason to conclude that you are not an intelligent and sensible person who cares about their health. It is your behaviour that is at fault here and not the beliefs themselves that are causing your cognitive dissonance, but this is not always the case.

Sometimes one, or both, of the beliefs you hold are faulty or just down right incorrect and often your behaviour is likely to be faulty or incorrect as a result. As such by removing the incorrect belief you also remove the incorrect behaviour. An example of this, which we can unfortunately pull straight from the newspaper headlines, has to do with the increasing number of people who

217

choose not to vaccinate their children. Here the conflicting cognitions could be as such:

> Vaccinations are an effective way to prevent many childhood illnesses

> Vaccinations are one of the main causes of autism in children

These two conflicting beliefs are likely to cause any loving parent severe mental discomfort. They want to do the right thing by their kids but there doesn't seem to be a great option here. As such they may decide upon the negative action of not getting their child vaccinated at all.

However, by re-examining the conflicting beliefs they may discover that one of them is flawed. In this case all of the scientific evidence points to the conclusion that childhood vaccinations do not cause autism.[*] In fact, the most influential paper in support of a vaccination/autism link was officially retracted by The Lancet[†], the World's leading medical journal, in 2010 due to part of it

---

[*] I'm not saying here that vaccinations are completely risk free, no medical procedure is, just that the evidence supports the conclusion that they do not cause autism. That said vaccinations do have a proven track record of reducing the chance of contracting many potentially fatal diseases. Speak to your doctor if you have any concerns.

[†] Andrew Wakefield's 1998 paper was official retracted from *The Lancet* on February 2, 2010 on the grounds that elements of the manuscript had been falsified. Additionally on January 28, 2010, the British General Medical Council ruled on a six year investigation stating that Wakefield had "failed in his duties as a responsible consultant", had acted against the interests of his patients, and had acted "dishonestly and irresponsibly" with regards to his controversial research into the link between the MMR vaccine and autism. As such the scientific support for the idea that vaccines cause autism effectively no longer exists.

having been falsified. Upon learning this the parent may abandon their preciously held belief that vaccinations cause autism and as a result take steps to correct the faulty behaviour that resulted from this. Once again this dissonance will disappear.

Forgetting or reducing the importance of one or both of the conflicting ideas

With this approach there are two options open to our smoker. Firstly they could acknowledge that, in this situation at least, maybe they are not being intelligent or sensible or they could even decide that they don't really care about their health that much. This effectively removes one of the conflicting ideas and as it takes two to tango the cognitive dissonance goes away as well. The other alternative is to just forget the fact that smoking is bad for you or at least conclude that it is not that big of an issue. Again this would remove one of the conflicting beliefs and once again the problem is resolved. The smoker's behaviour remains the same, they carry on smoking, but the conflicting beliefs are no longer a problem.

Now while this may sound similar to changing your beliefs there is an important difference. With this method of resolution neither of the starting beliefs actually changes. You still think smoking is bad for you and you also still sensibly care about your health. However, now one or both of these beliefs are no longer really of concern to you. As such the dissonance is probably still there, just hidden away or at a level you can manage, but the possibility remains that it may come up again as you have not really dealt with your problem, you are simply ignoring it.

Justifying our behaviour by introducing a new idea to the mix

As I said before this is where things generally start to go wrong. Rather than changing the behaviour that you know to be wrong or re-examining the underlying beliefs you hold to see which, if any, of them need to be changed, this method of resolving cognitive dissonance involves introducing a new idea in order to smooth over or justify the holding of two conflicting ideas.

Now there are occasions when this can actually work out, such as when two ideas appear to conflict because you are actually missing that vital piece of evidence that allows them to gel together. However what happens more often than not is that the new idea is just as invalid as the behaviour or false ideas that triggered the dissonance in the first place. In the example of our smoker this sort of dissonance resolution might work out like this:

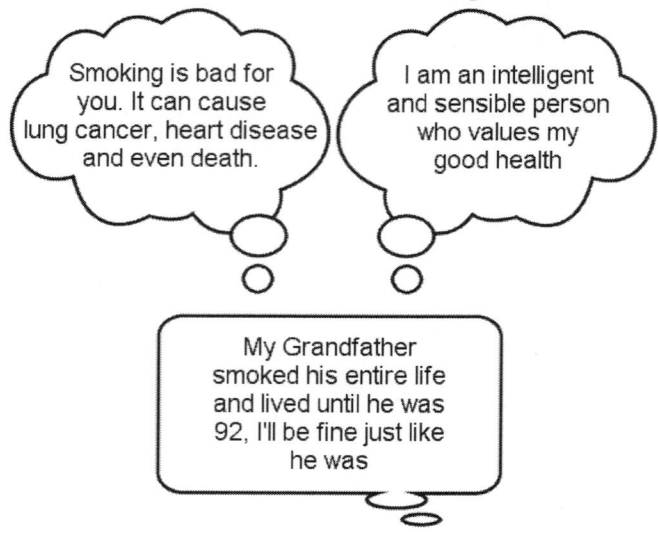

> Smoking is bad for you. It can cause lung cancer, heart disease and even death.

> I am an intelligent and sensible person who values my good health

> My Grandfather smoked his entire life and lived until he was 92, I'll be fine just like he was

This new idea may be entirely accurate, your Grandfather may well have been lucky enough to escape any ill effects from a lifetime of smoking, and as such this may help to resolve your cognitive dissonance. You still accept that smoking is bad for you and you still care about your health, but other members of your family have smoked and had no health issues because of it and so you conclude that you have nothing to worry about either. Alternatively you may just decide that you enjoy smoking too much to quit or that it keeps you thin and that this is more important to you. These approaches may set your mind at ease, and this sort of reasoning is common amongst smokers, but that doesn't necessarily make your conclusion any less fallacious.

So let's bring this back to skepticism shall we? As we covered at the beginning of the chapter cognitive dissonance is strongest when it involves deeply held beliefs such as, though not exclusively, the beliefs we hold about the type of person we are. The stronger the belief then the stronger the dissonance when something comes into conflict with that belief. The stronger the dissonance the more inclined we are to try and resolve it one way or another. On top of this, as we learnt with confirmation bias, people are more inclined to favour their existing beliefs and so dismiss, diminish or twist new evidence against them in favour of what they already hold to be true, rather than admit that they were wrong and so abandon that belief.

American social psychologist Leon Festinger, the man responsible for developing what is now known as the Theory of Cognitive Dissonance, talked about this very issue in his 1956 book, When Prophecy Fails. The book examined the members of a UFO doomsday cult and the effect that a failed end of the World prophecy made by their leader, which included the rescue of the cult members by aliens, had on their beliefs. To his surprise Festinger discovered that the failure of the World to end actually increased the commitment of the cult members to their leader and resulted in more fervent proselytising of their belief system.

Festinger concluded that this was the result of them trying to resolve the conflict between two cognitions, those being "the leader said the World would end" and "the World didn't end", by introducing a new idea into the mix. Rather than accept that their leader had been mistaken and, as an extension of that, that they themselves had been in the wrong, the cult members concluded that their faith had caused the aliens to save the entire planet instead of just the members of the cult. As such the failure of the World to end actually became a reason to believe that their leader was telling the truth rather than an argument against it.

Of course there is a nasty flip side to this coin. Cult members may resolve the dissonance caused by the failure of the UFO to pick them up even though their leader said it would by blaming their own unwillingness to demonstrate their faith to the fullest extent. As such when their leader states that the UFO is going to give them a second chance they will be even more willing to take extreme measures and participate in mass suicide as a sign of true devotion. On a more individual level a battered wife may resolve the dissonance caused by her husband saying he loves her and him hitting her by coming to the conclusion that she is to blame for his actions.

A somewhat famous example of this in action comes from 2011, when Christian radio host Harold Camping announced to the World that the Biblical Rapture[*] would occur on May 21st that year, prior to the destruction of the World on October 21st. The story was picked up by media outlets all over the World, spurred on by his believers as well as atheist, skeptic and religious detractors. When May 22nd arrived and the rapture had not

---

[*] The rapture, in case you are unaware of it, is the idea that the Christian faithful will be taken up into Heaven in order to spare them the seven years of suffering that they believe will take place before Jesus returns and the world as we know it ends.

occurred Camping stated that he was "flabbergasted" by this turn of events and "looking for answers". Well it didn't take him long to find them as by May 23rd he had reinterpreted his prophecy and announced that the rapture had actually taken place, only it had been a spiritual rapture. The physical rapture of the faithful, he added, would now occur on October 21st, at the same time as the prophesised destruction of the World. Now in case you were not paying attention on that day the World did not in fact end on October 21st 2011, once more proving Camping's prediction wrong. Perhaps unsurprisingly Camping chose to resolve his dissonance by sticking to his convictions about the end of the World and stating that he and his followers "humbly acknowledge we were wrong about the timing." He still believes the World will end any day now, but it keeping the date to himself this time round.

Now while this story is somewhat amusing it is important to note that things like this do have real world consequences. Camping spent millions promoting his belief that the rapture was coming, with most of the money coming from people who donated from their retirement and college funds in the belief they would never be needed. Some even quit their jobs as May 21st approached and one woman even attempted to kill herself and her children, so desperate was she to spare them the horrors she was convinced were coming. And sadly, Camping is not the only one out there making claims like this.[*]

---

[*] Whilst researching for this book I came across another, far less public, example of failed predictions being used to justify stronger belief. I discovered a website, now unfortunately no longer online, for an organisation called Rapture Ready that, much like Harold Camping, was predicting the Biblical rapture. In this case they believed it would take place on September 21st 2009. So on September 22nd 2009 I headed back to the website to find out what they had to say about us all still being here, only to find they had moved the date of the rapture to October 21st 2009, without a single mention of the fact they had been wrong the first

And this brings up another very important aspect of cognitive dissonance, that being that it can result in a whole load of self-deception. This is something a good skeptic should keep in mind before condemning the actions of a psychic or faith healer for example. It is possible that the person being most deceived by their act is the psychic themselves. As a result of cognitive dissonance and confirmation bias the psychic could truly believe that they have the ability to communicate with the dead. And, even if they know it is all an act, cognitive dissonance can result in them convincing themselves that this doesn't matter, after all they are helping people, aren't they?

This also helps address that age old question of how do bad people sleep at night. The answer is simple, they just resolve the dissonance between the belief that they are a good person and the evidence that this is far from the case by dismissing the evidence or introducing a new idea to justify their actions. The vast majority of bad people do not view themselves as bad people; the vast majority of those who joined the Nazi party honestly believed that they were doing what was best for their country, they really thought that they were the good guys. Most people would not join an organisation that proclaimed itself to be evil, no matter how evil the person in question may be. In their mind they are always on the side of the angels.

You need look no further than magician James Randi's Million Dollar Challenge to see this sort of self-deception in action. As the name suggests the Million Dollar Challenge promises to pay out the sum of a million dollars to anyone who can demonstrate

---

time. October 21st came and went and they moved the date yet again. In fact by the time I gave up checking they had moved the date of the rapture a further ten times, with the last time said to be before Rosh Hashanah, the Jewish New Year, 2010. At no point did these continuous failures have any effect on their apparent conviction that they were right and that the rapture was just around the corner.

Building your Skeptical Toolkit

psychic or paranormal abilities under mutually agreed upon test conditions. Many people have applied to be tested over the years but as of yet the money remains safe in the bank with no one coming close to proving their claim.

So how do the psychics resolve the dissonance caused by the conflicting beliefs of "I have psychic powers" and "my powers completely failed under test conditions"? Well the vast majority of them do so by justifying their belief in their own powers by adding a new idea into the mix. They convince themselves that the test had to be rigged or that their powers can only be used to help people and not for monetary gain, though most still charge for their services. Perhaps most common is the belief that their powers cannot be tested and that therefore science itself is somehow to blame for their failure. In fact this last argument is popular in all areas of the paranormal and pseudoscience and is used to explain the failure of everything from acupuncture to dowsing, homeopathy to astrology when examined under test conditions. My claim simply can't be tested by science, which is why it failed to perform as promised under test conditions. Science is to blame here, not my claim.

But before you, as a budding skeptic, start to feel all superior and immune to the effects of cognitive dissonance it is important to remember that everyone suffers from cognitive dissonance from time to time. As far as we can tell it is a universal human condition and being a skeptic does not inoculate you against it. If they are not careful a skeptic may find themselves automatically dismissing compelling evidence for some paranormal or pseudoscientific claim simply because to accept its legitimacy would require them to admit that they were wrong about something they strongly believe to be true.

A good skeptic must always keep this in mind and be willing to re-examine the beliefs they hold in the light of new evidence. After all our aim should always be to believe as many true things and as few false things as possible. As such a good skeptic should be pleased when they are shown to be wrong about something. That's one more false belief out of the way and one more new idea to examine by the light of reason.

One last thing before we move on to the next chapter. What should you do when you identify the effects of cognitive dissonance in others? Well depending on the situation there are many different things you can try to help them see the flaws in their reasoning or the incorrect way that they have resolved a conflict of ideas. As such it is probably easier if I just tell you the one thing you shouldn't do, and that is tell them that they are wrong.

Now this is a lot easier said than done. I myself do it all the time, in fact I have probably done so in the book a few times, but the evidence shows that telling someone they are wrong will simply increase their dissonance and, as a result, their flawed response to it. After all no one likes to be told that they are wrong. It is a much better approach to acknowledge their point of view and, if it is appropriate to do so, explain why you personally do not agree. You may even want to admit that you are open to having your mind changed on the matter. Now it is true that this approach won't win you any instant converts, but at least it shouldn't further cement their convictions and you never know what seeds you may have sown.

I Remember When
Past lives, false memories and Satan!

So now let's once again take things another step further. We've looked at how confirmation bias and cognitive dissonance can have a dramatic effect on the things we believe and how, if we let them run amok, they can result in us believing all manner of untrue things. But so far everything we have looked at has had a foundation in real ideas that we've had and upon real evidence that, while we may have interpreted it in a biased fashion, does actually exist. But what if the ideas and memories that we have, and upon which we build our beliefs, are themselves false? What if that memory you have, the one that forms the core of your belief in some paranormal claim, is of something that never actually happened?

False memories are actually surprisingly common and most people have a few memories that are in fact false, though they are most likely completely unaware of this. So what exactly is a false memory? Well very basically it is a memory we have that is wholly or partially incorrect in some way. It may be that we are simply misremembering an event that actually did happen or mixing up memories of things that happened at different times and remembering them as a single event. It can be the case that we confuse things that only happened in a dream with memories of an actual event or even that we imagine something happening and later remember it as though it really took place.

Your Brain Hates You!

This may come as a surprise to some people, and surveys show that around 27% of us believe this,[*] but our memories do not act like video cameras and we do not keep perfect records of the things we experience, something that 36% of us think is the case. Dr Chris French, a psychologist at Goldsmiths College in London specialising in Anomalistic Psychology Research[†], likes to use the following example to explain how our memories actually work. Imagine a clock face upon which the numbers are represented by Roman numerals. How is the number four displayed? Is it "IV" or "IIII"? Think about it for a moment until you are happy with your answer. What Dr French has found from asking this question many times over the years is that the vast majority of people will confidently proclaim that the answer is "IV", and they would be wrong. On most clocks that use Roman numerals, though there are of course exceptions to every rule, the number four is displayed as "IIII". Don't believe me? Well do a quick Google image search and you will see that I'm right.

So why do most of us get this wrong? Well when asked to recall this detail most people will try and produce a mental image of what they think a clock face using Roman numerals should look like. We then fill in the gaps in our actual memories with the knowledge of how Roman numerals usually work and the expectation that this applies to clocks as well. As such what we believe is a memory of looking at a clock face is actually a reconstruction based upon general knowledge, beliefs, expectations and actual recollections, as well as snippets from dreams, fantasies and our ever-active imaginations. And this, for the most part, is how all our "memories" work.

---

[*] Somewhat worryingly a more recent survey carried out in 2011 found that 63% of the U.S. population hold the belief that memory acts like a video camera, with 48% thinking our memories are permanent.
[†] The study of why people believe unusual things.

Building your Skeptical Toolkit

It is also not uncommon for people to insert themselves into the memories of others. Perhaps you have a friend or family member who has recounted a childhood story so many times that you now remember it as if you were there, or that the event actually happened to you rather than them, when in fact you were somewhere else entirely at the time of the incident.

An example of this last type of false memory from my personal life involves a work colleague and an amusing phone call. The call came in to me at the service desk on which I worked at the time. The caller was phoning to report black smoke coming from under one of the doors in the office. I immediately told them to sound the fire alarm and evacuate the building and that we would get someone over to investigate. The caller told me that she couldn't do this and could we send someone over to do it for her. I tried to convince her that she really needed to pull the alarm but she was adamant that this was not something she could do. Finally I asked her why she couldn't sound the alarm. "Well," she said "we have just moved into the building and we haven't worked out whose responsibility it is to do that yet." Needless to say I was gob smacked by this and it actually took my manager to convince her to sound the alarm and evacuate what was quite likely a burning building.

All of which brings me to my point. I recounted this story a number of times over the years. It made for a good tale to tell new employees when they enquired as to what sort of calls we had to deal with. Anyway about a year or so after this had happened I came into the office to hear the story being told once again, only with a slight difference. The lady telling the story was recalling the event as though she had taken the call herself, and this wasn't a one-time thing or just a way to improve the narrative of the tale. In her mind this event happened pretty much exactly

229

as I recall it, but with her as the central character rather than me. This is a very typical kind of false memory.

Of course the issue can become compounded if two people's memories start playing tricks on them. Maybe you have a memory of a childhood event in which you are convinced that your best friend was involved. You recount the event to your friend and they, not really remembering it themselves, take you at your word and fill in the blanks in their own memory with details from similar events that the two of you did in fact share. The fact that they were out of the country at the time of the event you are remembering no longer comes into it.

I have little doubt that many of my own childhood memories are false, constructed from stories my parents have told and my own imagination rather that real memories. In fact if any of you have memories from before the age of three then you can pretty much be sure that they are false. The left inferior prefrontal lobe, the area of the brain responsible for storing long term memories, is undeveloped in infants and as such the brain function for encoding and classifying memories simply doesn't happen before three years of age. But while this all sounds fairly harmless and maybe a little amusing there is a profoundly negative and dangerous side to false memories. Just ask Dr Donald Thompson, himself somewhat ironically an expert in memory, who was arrested for rape as a result of false memories. Just before the attack Dr Thompson was conducting a live interview on television, a program that the victim saw and as a result "*apparently confused her memory of him from the television screen with her memory of the rapist*". It is easy to see how our minds play tricks on us but so far we have ignored the fact that false memories can very easily be implanted, either intentionally or accidentally, in our heads by others.

Building your Skeptical Toolkit

But surely it can't be that easy to get someone to believe something that never happened. Well you would be surprised. One of the earlier experiment into false memories involved showing people simulated crimes or accidents. Later on the participants were fed incorrect information about the event, such as the man involved had curly rather than straight hair. When interviewed on what they had seen many of the people claimed to have seen a man with curly hair.

In a later experiment undergrads were shown a short video of two cars driving along, one a BMW and the other a smaller, less powerful Volkswagen Polo. In the video both vehicles were travelling at the same speed, however when questioned the very next day participants remembered the BMW travelling significantly faster than the Polo. Another experiment involved even shorter periods of time between the event and the generation of a false memory. Professor Richard Wiseman conducted an experiment in which people where shown a video of an alleged psychic apparently bending a key with his mind, after which he placed the key on the table and said *"If you look closely you can see it's still bending."* Of course the key did nothing of the sort, and yet 40% of the people who heard this simple verbal suggestion believed that they saw the key continue to bend. When this suggestion was removed virtually no one reported seeing the key continue to bend. Dr Chris French then took Wiseman's research one step further. Using the same video he conducted the experiment again, only this time he had an accomplice pretend to also see the key continue to bend. When the verbal suggestion was combined with the agreement of a co-witness the number of people who reported that they saw the key continue to bend rose to 60%.

In fact you may not even be able to fully trust your memories of things that just happened to you. Research shows that the distorting effects on memory can occur within seconds of the event. In 2002 such memory problems caused issues for those investigating the sniper attacks in Washington DC. A number of witnesses reported seeing a white truck or van fleeing from the scene of several of the attacks thus influencing the investigation in this direction. When the snipers were eventually caught they were driving a blue car and apparently had no connection to a white van what so ever. So why were so many witnesses convinced that they has seen a white van fleeing the scene? Well Elizabeth Loftus, a psychologist at the University of California, believes that we may have witnessed a case of nationwide memory contamination. During one of the earlier attacks it was reported that a white van might have been spotted in the area and this piece of information was subsequently repeated by the media in almost all reports on the subject. As such potential witnesses were already primed to be on the lookout for a white van and unsurprisingly, when the next attack occurred, this is what people remembered seeing.

Even so called "flashbulb memories" are not as reliable as many of us believe. A flashbulb memory is a memory of a particularly emotional or significant event that we appear to remember with almost photographic clarity. If I were to ask you where you were and what you were doing on September 11th, 2001 I am sure you would have no trouble telling me.[*] Likewise you all probably have a clear picture of what you were up to when you first heard the news of Princess Diana's death in 1997.[†] But how accurate are these memories?

---

[*] I was at home sick in bed.
[†] I was going to get myself some breakfast and was told by my parents sitting at the dining room table. Or at least this is how I remember both of these events.

Building your Skeptical Toolkit

A study was carried on a specific flashbulb memory related to the 1986 *Challenger* space shuttle disaster. The subjects were asked to recall where they were and what they were doing when they first heard the news that the shuttle had exploded a minute after take-off. The first time they were asked was just 24 hours after the event and the second was two and a half years later. Here is one set of descriptions from those recorded:

Description 1.
"I was in my religious class and some people walked in and started talking about (it). I didn't know any details except that it had exploded and the schoolteacher's students had all been watching which I thought was so sad. Then after class I went to my room and watched the TV program talking about it and I got all the details from that."

Description 2.
"When I first heard about the explosion I was sitting in my freshman dorm room with my roommate and we were watching TV. It came on a news flash and we were both totally shocked. I was really upset and I went upstairs to talk to a friend of mine and then I called my parents."

Clearly these two descriptions are significantly different and the psychologists carrying out this experiment reported that a third of the subjects questioned gave similarly differentiated descriptions. But before we get to the really dark side of false memories let's look at the grey middle ground and how false memories have an important role to play in many areas of the paranormal and pseudoscience. Back in 2008 British actor and TV presenter Tony Robinson hosted a program in which he investigated alleged paranormal events said to have taken place in and around the UK.

Your Brain Hates You!

One program focused on the subject of past lives and in particular a series of events from the 1960's during which numerous unconnected people recounted past life dreams of being members of a persecuted 13th century religious sect from France, known as the Cathars, to a psychiatrist in Bath. Part of his investigation involved visiting a past lives regression therapist and experiencing first-hand the sort of thing she did.

The therapist got him to lay back, close his eyes and relax and set about placing him in a light hypnotic state. She then asked him to picture himself getting younger, slowly taking him back through the years, through early adulthood, through his teens, through childhood and then infancy. She then asked him to picture a time before he was born and to imagine himself still going back even further. "*And now,*" she said in her soft, rhythmic tone, "*you will find yourself emerging into your past life.*" It was at this point I literally yelled at the telly. Without maybe realising it the therapist had just completely invalidated the experiment. Rather than simply allowing him to take the journey wherever it may lead she had placed him in a highly suggestible state and then *told him* that he had lived a past life.

From this point on things went from bad to worse. She continued to ask him leading questions, telling him exactly what details he needed to recall to construct the "memory". Is it hot or cold? Are you standing up or sitting down? Are there other people around or are you on your own? All questions were presented as an either/or choice, never once did she leave it open for him to fill in the gaps on his own. Unsurprisingly, under this guided process, Tony was able to recall in great detail an event from his "past life" as a soldier in India.[*] Is it possible that he really was experiencing

---

[*] It is worth noting for those who are unaware that at the time Tony Robinson also presented a program on archaeology called Time Team on Channel 4 as well as a number of other documentaries on historical

234

Building your Skeptical Toolkit

memories from another life? Well as good skeptics we have to at least consider that possibility, however it is far more likely that, in this instance at least, his memories were entirely false and constructed thanks to his imagination and leading questions from the therapist.

Those who work in the law enforcement community are all too aware of the dangers of leading questions. If they are not careful the questions asked by a detective can influence the answers given by a witness so that they "remember" details that they believe the detectives want to know rather than the actual things they saw. Again this is something we all do from time to time. Most of us like to be helpful and when we can see someone is expecting something of us we are inclined to give them the answer we think they are looking for, even if it is not accurate. The problem is that we can often fool ourselves into believing that the answer we provided was in fact the right one.

You can all picture the scene. The police have a witness to some crime and are trying to get them to identify the perp. They lay out a series of photos on the table; the witness scans them and quickly identifies old man Withers who runs the haunted amusement arcade as the person responsible. Wait. I may be getting things mixed up with Scooby Doo and old cop shows. The thing is that in real life it is rarely done this way.

The police understand that witnesses often feel under pressure to provide them with the information they need or just really want to help them catch the criminal. As such when presented with a series of photos a very natural inclination is to assume that the guilty party is amongst the faces before you. Our urge to help and our fallible memories can cause us to make an identification when in fact the perpetrator is not laid out before us at all. The police

---

events. His ability to construct believable sounding events from history therefore can probably be explained without recourse to past lives.

are well aware that something like this can result in false memories just as easily as directly leading questions. As such, it is not uncommon for them to start by presenting a series of photos of people they know where not involved in the crime in order to try and eliminate the false positives brought on by poor memory and overly enthusiastic witnesses. Additional, as we saw with Dr Thompson, it is not uncommon for witnesses to take a face they know from some situation completely unrelated to the crime and transpose it into the narrative they tell the police.

Let me show you how easily it is to implant a memory into someone's head, in this case your head. The following experiment utilises something known as the Deese-Roediger-McDermott (DRM) paradigm, named after the three psychologists who first developed the technique, and shows just how susceptible to false memories our brains really are.[*] Now this works best if you have someone read the words in the list below to you, with a second pause between each word, however as that may not be possible right now simply read the list to yourself and then turn the book over, though it may not work as well that way. You'll also need a pen and some paper for this. Here are the, slightly modified, instructions used in the experiment.

> Please listen carefully to the list of words. Do not write anything while listening to the word list. At the end of the list the reader will count to ten and then say the word "Recall." At that point you will have two minutes to try to remember and write down as many of the words as you can. You should start by trying to remember the last few words on the list,

---

[*] This exact example may not work on everyone. In the actual experiments numerous word lists were used. The specific word list in this example is taken from the paper *False Recall Does Not Increase When Words are Presented in a Gender-Congruent Voice* by Kreiner, Price, Gross and Appleby of Central Missouri State University.

Building your Skeptical Toolkit

but it is okay to write the words in any order. Do not worry about spelling.

Here are the words:

butter, food, eat, sandwich, rye, jam, milk, flour, jelly, dough, crust, slice, wine, loaf, toast.

Ok so if you are doing this alone stop reading now, turn the book over and place it to one side while you try to remember the words on the list.

Right, you can start reading again now...though you won't know that until you have already started to do so, hmmm. Anyway now compare the words you have written down to the words in the original list. Chances are that your list will include some of the following words: wheat, water, roll, piece, knife, cheese and most likely the word "bread". None of these words appears in the original list and yet the majority of people remember them anyway. This is because the list was designed around a critical lure, in this case the word bread. All the words in the list imply the word used as the critical lure and even though that word itself is not used you remember it being there.

If questions, either asked by a detective or a past life regression therapist, are not worded carefully it is very easy to imply the answer you are looking for without actually mentioning it. An exaggerated example of this may go like this. A detective is interviewing a witness and asks, "As your attacker loomed over you could you tell if he was tall or short?" Now I don't know about you but I have great difficulty imagining a short person "looming" over anyone. The wording of the question implies that the attacker was tall without actually saying so. This can result in the witness with a fuzzy memory giving the answers the detective is expecting rather than an accurate account of what happened. The same applies to well-meaning psychotherapists who, in attempt to help their patient, may suggest something along the lines of *"you may not remember being sexually abused but you do appear to have the symptoms. Do you have any idea who may have done this to you?"*

Children in particular are very susceptible to leading questions and implanted memories. One set of experiments showed that detailed and entirely false childhood memories of things ranging from being lost in a supermarket all the way to surviving a vicious animal attack can be implanted utilizing little more than a trusted friend or relative backing up the things the psychologist is saying. Thrown in a few doctored photos or fake adverts and you can have people believing they took hot air balloon rides that never happened or that they shook hands with Bug Bunny at Disneyland.*

An example of the type of image used to induce false memories.

---

* In case it is not immediately clear to you Bug Bunny is a Warner Bros character and so most definitely would not have been at Disneyland. And yet using just a few false advertisements showing Bugs Bunny at Disneyland psychologists were able to convince 62% of those involved that they had shaken his hand and 46% that they had hugged him.

Building your Skeptical Toolkit

And this brings us back to the dark side of false memories. But first, if you will excuse me the indulgence of another analogy, let me tell you a little story about my niece.

When she was around two years old my niece and my sister came to visit me. Now I'd seen my niece just a few days earlier and for some reason assumed that she'd had her hair cut during that time.* So I asked her if she'd had her hair cut to which she happily informed me that she had. I then asked if she had gone to the hairdressers and again she said she had. I proceeded to ask her a few more questions about having her hair cut and she continued to answer in the affirmative, even going so far as to fill in some details of her own and act out cutting her hair with her fingers. When she ran off to play I looked up to see my sister looking at me with a puzzled look on her face. *"Well,"* she said, *"I have no idea what that was all about. We haven't cut her hair in over a month and we have never been to the hairdressers to do it."*

My mistaken assumption and my niece's imagination, perhaps mixed with her trust that her uncle knew what he was talking about, had resulted in an entirely false memory of an event that never happened. Now in this example there was no harm done, but what would have happened if my assumption had been far less innocent and my niece's natural willingness to play along far more incriminating?

While all of us suffer from false memories from time to time there is a real difference between this and what has become known as False Memory Syndrome.† Dr. John F. Kihlstrom, professor of

---

* Clearly my own memory was playing tricks on me.
† I really feel that I should add a quick caveat here. False Memory Syndrome, or FMS, is something of a controversial diagnosis and currently has not been recognised as a valid condition in the Diagnostic and Statistical Manual of Mental Disorders. I mention it here because it provides an effective short hand for describing a number of complex and interconnected psychological conditions that are recognised by

psychology at the University of California in San Francisco, describes False Memory Syndrome as "*[A] condition in which a person's identity and interpersonal relationships are centered around a memory of traumatic experience which is objectively false but in which the person strongly believes.*"

Picture someone you know, someone younger than you whom you have known for most of their lives. It could be someone related to you or the child of a friend. Now imagine this person goes into therapy for help with something normal and every day, such as anxiety, depression or good old fashion stress, and that after a few sessions they suddenly accuse you of sexually abusing them when they were younger. Now you have no idea what they are talking about, you never did the things they accuse you of, and yet they are adamant that you did. You can easily see how something like this can destroy the lives of both those accused and their accusers.

Starting in the 1980s a moral panic swept over the United States and subsequently many other parts of the World. In therapy sessions across the country, and with the help of such techniques as hypnotism, an alarmingly large number of people were recovering memories of physical and sexual abuse at the hands of those engaged in satanic and occult rituals. At its height, these reports encompassed a worldwide conspiracy of the rich and powerful and involved stories of children being abducted or even specifically bred for ritual sacrifice, pornography and prostitution.

---

mainstream psychology. Additionally Dr Kihlstrom's description of FMS makes it the perfect term to use when talking about the negative effect that false memories can have on people, something that undoubtedly does happen, regardless of the conditions actual validity. For more information on FMS check out the website of the British False Memory Society at www.bfms.org.uk. For a counter point psychologist Ken Pope has a number of papers on his website that discuss some of the criticisms against FMS. http://kspope.com/memory/memory.php

Building your Skeptical Toolkit

The problem was that, aside from these "eye-witness" accounts, there was no evidence to support the existence of a worldwide satanic cabal of baby killers. Sadly this fact did little to help those convinced of horrific crimes that never happened or reconcile those families torn apart by fictitious memories of childhood abuse by masked figures in the basement of the family home. So what was going on here? Well the problem, at least in part, seemed to stem from the use of dubious therapeutic techniques, the belief in the inherency of memory and the acceptance of the idea of repressed memories. We have already looked at how unreliable our memories can be about, well, almost anything, so let's focus on the other two ingredients that play a role in False Memory Syndrome.

We all know about the idea of repressed or recovered memories. We've all read books or seen movies where someone has repressed the memory of some terrible event only for them to remember what actually happened in the nick of time to defeat the bad guy. Most of us have come to accept the idea that if an event is bad enough then our brains will try to protect us from it, pushing it deep down into our subconscious where it reveals itself only in bouts of insomnia and excessive drinking. This belief is even common amongst psychotherapists with a 1994 survey showing that 60% of those questioned believe that repression was a major cause of people forgetting and around 40% believing that people were unable to remember much of their childhood due to them repressing traumatic events. What most people don't know is that this Freudian idea of repressed memories has very little evidence in its favour. In fact what the evidence does seem to indicate is that people who suffer traumatic events such as sexual abuse actually tend to have difficulty thinking about anything else. As for the idea that hypnotism is an effective way of recovering or enhancing memory this is another popular but incorrect belief, with between 70-90% of psychology students believing this to be the case. We have already looked at how easy it is to create false memories under hypnotism but there is more to it than just this. In addition to an increase in false memories hypnotism also tends to increase the effect of memory hardening. Memory hardening is where someone's confidence in the accuracy of a memory increases. This is generally believed to be the result of people's mistaken confidence in hypnotism's ability to recover accurate memories. As such any memories they do recall under hypnosis are automatically believed to be more accurate and reliable than memories recalled under more natural circumstances. Because of this many states in America have banned the testimony of witnesses who have been hypnotised from being admissible in the courtroom.

As Dr French speculated when writing about the link between sexual abuse cases and false memories *"should it not give us pause for thought that exactly the same "memory recovery" techniques, including hypnotic regression and guided imagery, can give rise to apparent memories of being taken on board spaceships and medically examined by aliens, or a former incarnation as Napoleon?"*

We all suffer from false memories to some extent or another. Even those with so-called "photographic memories" are not entirely unaffected by memory reconstruction and errors. When someone insists that they have a very clear memory of being abducted by aliens they are probably telling you the truth. They really do remember such a thing happening to them; however that doesn't mean that the memory is real.

Skeptical Sidebar #4
The amazing memory trick

Seeing we have been looking at ways in which our memories can let us down I thought it would be fun to look at one way in which you can improve your memory, get one over on your friends and maybe get a free drink in the process. Here's what you need to do. Next time you are out at the pub with some friends propose a challenge to one of them. If they can remember more words from a list of twenty items than you can then you will buy the next round, if not then it is their shout. Once they accept have the rest of your friends come up with a list of twenty items and write them down on a piece of paper and then place it on the table/bar where both you and your challenger can see it. Have the writer give you no more than a couple of minutes to memorize the list and then have them remove it from the table and hide it away safely in a pocket or purse. Then ask the people at the table what they would like to drink so that you and your challenger both

Your Brain Hates You!

know what you have to buy should you lose.* Now let your friend recall as many words as they can, chances are they will manage around 7 or so, and then when it is your turn flawlessly recite all twenty words on the list...then do it backwards just for fun. Free drink time, easy.

What? Oh yeah, sorry there is a little more to it than that. How exactly do you go about remembering all twenty words so easily when your friend struggled to remember less than half of them? Well here's the trick, you use a memory aid called the mnemonic link system. Here's how it works. Let's say this is the list of twenty words your friend comes up with.

1. Rabbit
2. Bucket
3. Orange
4. Hat
5. Door
6. Bank
7. Mirror
8. Book
9. Car
10. Biscuit
11. Telephone
12. Fridge
13. Sandwich
14. Mouse
15. Coat
16. Bus
17. Banana
18. Computer
19. Church
20. Toilet

Before I tell you how you can remember the whole list, give it a go on your own. Look at the list for no more than a minute or so and then close the book and see how many you can remember. Most likely you will remember what cognitive psychologist George A. Miller called the "The Magical Number Seven, Plus or Minus Two", that is you will remember around seven items from the list, give

---

* This is actually a way of distracting your challenger and making it harder for them to just go over and over the list in their minds. Sneaky, but you want to win don't you?

Building your Skeptical Toolkit

or take a couple. Miller found that most people can only hold around seven "chunks" of information in their working memory at any one time. These chunks can be simple such as single numbers or letters or more complex such as groups of numbers, words or even phrases. You can test this idea out for yourself. Here is a list of twelve numbers, have a quick read through it and then close the book and see how many you can remember.

    4   8   3   5   9   7   4   2   0   6   3   8

Most likely you were able to remember the first few fairly easily but then started to struggle a bit and by the time you were half way through the list you had pretty much reached your limit. This is because there are simply too many chunks of information for you to deal with. But now let's try something; let's take the same 12 numbers and group into four blocks of three.

    483   597   420   638

Now read through the list, close the book and see how many you can remember this time. It is very likely that you were able to remember all twelve numbers with relative ease this time round, so what changed? Well by grouping the numbers this way we reduced the number of chunks of information you needed to remember from twelve to four. Four items is well within your working memories limit of seven plus or minus two and so you are able to remember all the numbers this time round.

Ok, so this explains why you could only remember a few items from the list, how do you go about remembering them all? As mentioned you do this by using the mnemonic link system. A mnemonic is basically a device that aids memory. There are numerous different types of mnemonic, however the one we are looking at here involves creating mental links between the items on the list and plays on the fact that generally the more complex, and unusual, a memory is the easier it is to remember.

Let's go back to our list and start making links shall we. The way to do this is to create a mental picture that links one item on the list to the next one, and the more unusual that mental picture the better. So the first item on our list is a rabbit and the next is a bucket. Well you might create a mental picture of a rabbit sitting in, or better yet stuck in a bucket. Next we need to link the word bucket to the word orange. You might do this by picturing a bucket full of oranges or maybe an orange being used as a bucket. The next word on the list is hat and we need to link this to the word orange. Easy, just picture an orange wearing a hat. Work your way through the whole list creating these mental pictures. You might picture a telephone in a fridge or a mouse wearing a coat for example. Take a couple of seconds to really picture each link in your head before moving onto the next one. The best mental pictures are ones that you can engage with on more than one level. So with the rabbit in the bucket for example you might imagine how uncomfortable the rabbit is and think about how you would feel to be stuck in a bucket. Alternatively you might focus on the humour of the situations, even allowing yourself a little smile as you picture the silliness of the situations. Both of these approaches will create a stronger memory than the picture alone, so try to come up with mental pictures that you have strong feelings towards one way or another.

Once you have created mental pictures for all the items on the whole list close the book once more and starting at the beginning work your way through the whole thing. You should find that the mental pictures spring easily to mind and lead naturally on to the next in line. If you find yourself getting stuck at any point it might be because your mental picture for that link isn't that strong. If this happens just try coming up with something else and give it another go. Once you have gone through the whole list try doing it backwards. Now you didn't actually learn it this way but you should find that the mental pictures work just as well in reverse.

Once you have mastered this it is time to head to the bar and start winning free drinks off your mates...or you could use this and other mnemonic techniques to do something useful, like revise for an exam or remember the names of people at a party.

Your Brain Hates You!

I am not a number...
The power of conformity and the wisdom of crowds

I want to take a couple of pages to expand on something I mentioned in passing in the previous chapter, and that's the idea that other people can directly influence the things we believe. Psychologist call this natural inclination for us to be influenced by other people Conformity and, like with many of the other things we have looked at, it is not entirely a bad thing. It is thanks to the effect of conformity that we don't have to worry, for the most part at least, about finding someone driving towards us on the wrong side of the road or walking out in front of us when we have a green light. Conformity also accounts for that most popular of British past times, queuing. We all know that when we are at the bank or the supermarket we have to join the back of the queue and wait our turn and we look unkindly at anyone who does otherwise. And there is nothing wrong with this. These are the things that keep our society ticking along nicely and stop it falling into anarchy.

But of course the problem with going along with the crowd is that sometimes the crowd is wrong. A good skeptic should remember that just because the majority of people accept a claim or idea doesn't mean that you should just accept it as well, or that the claim is right. Always evaluate the claim yourself and come to your own conclusion based upon the evidence. But then that may be easier said than done, as conformity is a powerful force to contend with.

In the 1950s psychologist Solomon Asch conducted a number of experiments into group conformity using student volunteers. A group of eight individuals were seated in a room and asked to look at a series of paired cards showing lines of various lengths. The first card in each group was the reference card and had a single vertical line on it. The other card was a comparison card and this showed three lines, labelled A, B and C, of various lengths, one of which matched the length of the line on the first card. All the participants had to do was state out loud which line on the second card matched the length of the line on the first card.

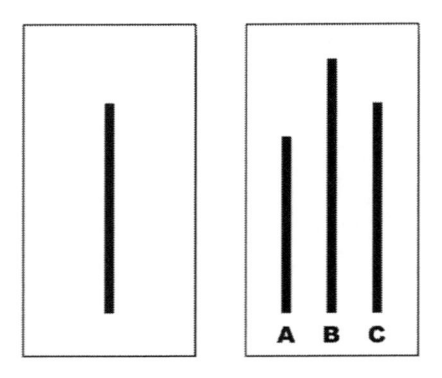

It all sounds easier enough but there was a hidden twist to the experiment. All but one of the people in each group was in on it with Asch and had been instructed to give set incorrect answers starting with the third pair of cards in the series. For example with the cards above the seven stooges would all state the line B was the matching line even though this is clearly not the case. Asch's experiment was designed to see how the remaining participant would react when confronted with the incorrect answers of the rest of the group.

When Asch conducted control experiments, where there was no pressure to conform to the opinions of the rest of the group, only one person out of the thirty-five tested gave an incorrect answer. However when placed in a group of people purposefully giving

Building your Skeptical Toolkit

incorrect answers almost 75% of those tested went along with the group and gave an incorrect answer on at least one set of cards. In fact 33% of those tested went along with the group's incorrect answers more than half of the time.

Asch also showed that the number of people in the group had an effect on the conformity of the subject being tested. If there was just one other person in the group then the subject would almost always make up their own mind regarding the cards no matter what answer the other person gave. The same goes for when there was just two other people in the group, however once the size of the rest of the group was over three the number of people conforming and giving incorrect answers rose significantly. When questioned later as to why they had given incorrect answers the majority of those tested admitted to thinking that the group was wrong but felt pressure to go along with them anyway. Some however actually believed that the group was giving the correct answers and that it was they themselves that were somehow mistaken, be it due to poor eyesight or misjudgement.

In order to test the idea that conformity can actually change people's perceptions psychologist George Berns conducted his own similar set of experiments. The main difference between his and Asch's experiment was that his subjects were wired up to an MRI machine during the process so that their brains could be studied to see which areas of it were involved in giving their answers.

A group of five people, one subject and four actors, were asked to look at two three dimensional shapes on a laptop screen and rotate them in their minds to see if the two shapes were the same. Before the actual experiment started the groups played through twenty practice rounds together and were encouraged to chat about the answers they gave. Then the group was split up and the subject was taken to the MRI room and the exercise was run for real. In a third of the tests the answers given by the other members of the group were not visible to the subject, in another third the answers were visible and the actors all gave incorrect answers while in the final third the answers were visible and the actors all gave correct answers. As with the experiments conducted by Asch the answers given by the rest of the group significantly influenced the answers given by the subject. The results showed the subjects agreeing with the incorrect answers given by the group an average of 41% of the time. But unlike Asch's experiment the MRI machine allowed Berns to see exactly what was going on in the subject's brains when they gave their answers.

Bern's hypothesised that if the subject's conformity to the group was the result of conscious decision making then they should see changes in the areas of the forebrain that deal with monitoring conflicts, planning and other higher-order mental activities. However if the conformity was due to an actual change in the subject's perception then they should see activity in the posterior area of the brain dedicated to vision and spatial perception. What Berns found was that when people went along with the wrong answers given by the group there was activity in the right intraparietal sulcus, an area devoted to spatial awareness, but no activity in the areas of the brain responsible for making conscious decisions. In short their perception of the objects before them was actually changed by the pressure to conform to the group. Interestingly Berns also found that those who went against the group experienced increased activity in the parts of the brain associated with emotional salience, thus supporting the commonly held belief that being a rebel makes you feel good.

What these two experiments show is that some people will go along with the group's opinion even when they personally think that it is incorrect. Also the influence of the majority opinion on an issue can actually change people's perceptions of the World around them. What this can mean of course is that the reality of the majority opinion itself can be a misconception as we have no way of telling how many people in the group actually hold the belief in question and how many of them disagree but are willingly going along with it anyway.

In addition to this, how popular we believe a belief is can be directly influenced by how familiar we are with that belief, as Kimberlee Weaver, a Postdoctoral Research Fellow at the University of Michigan, discovered. She found that hearing a claim or idea from the same person multiple times has the same effect on people with regards to how popular they think that idea is as hearing it once from multiple people. As such a vocal minority can convince you that their beliefs are in fact held by the majority simply by repeating them often enough. This may go some way to explain why fringe beliefs such as global warming denialism and the anti-vaccination movement get such a disproportionate amount of air time in the media when compared with how many people actually believe these things.

This effect is sometimes referred to as the Minority Influence and, as the name implies, is where the minority opinion influences the beliefs of the majority. Unfortunately, as with the two previously mentioned examples, this more often than not involved fringe pseudoscientific beliefs being blown up by the media in a way that implies that these ideas are in fact main stream. This in turn can result in misplaced conformity to an idea that isn't actually held by most of the group. A prime example of this is the MMR vaccination scare that rocked the UK in the early years of this century. One solitary paper by a now discredited doctor combined with the support of a small ideologically driven group of activists caused the media to flood the nation with untrue stories about the dangers of the MMR vaccine and, unsurprisingly, saw the vaccination rate for measles, mumps and rubella, as well as other diseases, plummet to dangerous levels all over the country.

Of course as a skeptic it is likely that you will find yourself opposing the majority point of view now and again and in these cases you need to know how to make minority influence work in your favour. Research shows that minority influence is most effective when it is presented in a clear and consistent way and supported by evidence that makes a strong, convincing case for your point of view. As such if you hope to change the majority opinion you need to make sure you present your argument in a simple and clear way and that you can back up the claims you make with evidence. Also make sure you are comfortable talking about the subject at hand as, unfortunately, people mistake confidence in a belief or idea for evidence that the belief or idea is accurate. Yeah I know, but what are you gonna do?

The final thing I want to cover in this chapter is really only tangentially related to the subject of conformity, but I can't really talk about the effect that other people have on our decision making without mentioning it, and that's something known as the wisdom of crowds. Very basically the wisdom of crowds is the idea that a large group of people are more likely to come up with an accurate answer to a question than one individual person on their own.

One common example of this is the frequently used fund raising technique employed by offices all over the Worlds, and that's the "guess how many sweets there are in the jar" game. People pay a pound to take a guess at how many sweets there are in a glass jar with the person making the closest guess winning the jar of sweets and the money going to charity. Now while it is completely possible for someone to guess exactly right more often than not if you take all of the guesses and work out the average this number will be more accurate than any individual guess. This is the wisdom of crowds at work.

Another example involves online, user driven encyclopaedias such as Wikipedia. Unlike most encyclopaedias where the information is static Wikipedia is open for anyone to edit the information as they see fit. Information that is seen to be incorrect is then itself edited until the entries represent the consensus opinion of the Wikipedia audience and, more often than not, therefore more accurately reflect reality with regards to the subject at hand.

Of course the wisdom of crowds only really works if the individuals that make up the group involved are offering up their own opinions, based upon their own examination of the evidence and independent sources of information, and are not simply conforming to the consensus of the group. With our sweet jar example if everyone comes up with a number that conforms to the previous guesses rather than making a guess based upon their own analysis of the evidence presented to them (that being examining the sweets in the jar for themselves) then the average will reflect that rather than how many sweets there really are in the jar.

The other important issue with the wisdom of crowds is to point out the situations where it doesn't work. In late 2009 English illusionist and mentalist Derren Brown presented a series of shows on Channel 4 entitled *The Events*, one of which involved him apparently predicting all six numbers of that night's National Lottery Draw. In a subsequent show he explained that he did this by utilising, amongst other things, the wisdom of crowds' effect, but alas for all of you wannabe lottery winners, Mr Brown was fibbing.

You see the wisdom of crowds only works when there is actually a specific answer to the question at hand. With the sweet jar example there is a specific, if unknown, number of sweets within the jar. Each person examines the jar and makes an educated guess, based upon some mental calculation involving the size of the sweets and the dimensions of the jar, as to how many sweets there are in the jar. Each individual's calculations will be accurate to a lesser or greater degree and therefore when an average of all the guesses is taken that number will represent the combined accuracy of all of these mental calculations.

With the lottery there is no evidence for the individuals to examine that has any effect upon the outcome of the next draw. Even if the number 12 has come up in every draw going back ten years that has no effect on whether the number 12 will come up in the subsequent draw or not. Any calculation, mental or otherwise, based upon the results of previous draws will in no way reflect the results of the upcoming lottery. As such taking all of these individual guesses and working out the average will not produce a more accurate guess than simply picking numbers at random. Derren Brown more than likely used clever camera tricks to pull off his prediction[*] and nothing more. Sorry.

---

[*] For more information on exactly how he did it, or at the very least the best guess as to how, check out Captain Disillusion's video on the subject over at www.youtube.com/user/CaptainDisillusion

Skeptical Sidebar #5
The gambler's fallacy and randomness

Let's say that the picture above represents nine flips of an ordinary ten pence piece, which way up do you think the coin is most likely to land on its next flip? In case you can't make it out from the picture the results currently run like this.

T   H   T   H   H   H   H   H   H

So do you think it is more likely for the coin to land on heads or tails on the next flip? Think about it for a moment, what does your gut tell you?

Ok times up. If your gut told you anything other than "I don't know, the odds are 50/50" then your gut is wrong and has just fallen foul of the gambler's fallacy. Allow me to explain.

The gambler's fallacy is the mistaken belief that, sticking with the example above, just because you got six heads in a row the coin must be due to come up tails and that therefore this outcome is more likely to happen. Other examples of this fallacy in action include a roulette wheel that has come up red the last twenty times prompting everyone to bet on black or believing that a slot machine that hasn't paid out in a while is due a jackpot. The fallacy occurs when people assume that this flip of the coin, spin of the wheel or pull of the handle is in some way connected to those that came before it and thus will be influenced by prior results, when in fact this is not the case.

This is because each event, each coin toss or wheel spin, is statistically independent, which means that each event has no statistical effect upon the event that comes after it, the two things are for all intents and purposes completely separate. If you flip an honest, non-weighted coin the odds of it coming down heads are 50/50, even if it has landed on tails the last ninety-nine times. This relates to the second mistake people make. They confuse the long term odds with short term results. Yes if you flip a coin a million times you will in all likelihood get an overall result that is pretty close to 50/50, however this does not mean that getting one hundred heads in a row is abnormal or in some way makes it more likely that you will get tails the next time. In fact the odds of getting one hundred heads in a row is exactly the same as getting any other combination of heads and tails from a hundred coin flips.

Take a look at the three coin flip sequences below. Which one would you say best reflects a random sequence of results?

Seq 1:	H	H	T	H	T	T	T	H	T	H
Seq 2:	H	T	H	T	H	T	H	T	H	T
Seq 3:	H	H	H	H	H	H	H	H	H	H

Now I am willing to bet that you went with the first sequence and you know what you would be right, well at least in part. The first sequence does represent a random sequence of coin flip results, but then so does the second sequence and the same for the third. If asked to write down a random sequence however it is most likely that you would produce something similar to the first sequence above. This is because people are bad at identifying random sequences. We assume that random means that the

Building your Skeptical Toolkit

values will alternate a lot and that there will be no repetitive patterns. As such when we look at truly random sequences, such as Pi[*], they tend to look wrong to us as they are often clumpy and have repeating sequences that we just don't expect to see.

This inability to correctly identify random sequences as well as the mistaken belief that if something hasn't happened in a while it is therefore due is what has kept casinos in business for years. In fact it is the reason the croupiers keep track of how often each number on a roulette wheel has come up and are all too happy to inform you that 20 black hasn't come up in the last three hundred spins of the wheel. They are hoping you will fall foul of the gambler's fallacy; it is how they make their money. We'll look more at our failings when it comes to numbers in the next chapter.

In the meantime, let's win you another drink shall we. As we have been talking about coins and gambling here is a little game you can play with your friends that makes the 50/50 odds of the coin flip work in your favour. Get a friend to pick a sequence of three possible outcomes from a coin toss. For example, they might pick heads, heads, tails. You then pick your own sequence, let's say tails, heads, heads. You then get someone to toss a coin and every time one of these sequences comes up the relevant person scores a point. The first person to ten points wins and the loser buys the next round.

Of course you're not foolish enough to leave things up to chance and instead you will be employing something that mathematician Walter Penney discovered in 1969. Penney, after whom this game is named, discovered that for each of the eight possible sequences of heads and tails available from three coin tosses

---

[*] You have an 86% chance of finding any 8 digit sequence in the first 200,000,000 digits of Pi, so why not see if you can find your date of birth in there - http://www.angio.net/pi/piquery.html

259

Your Brain Hates You!

there is another sequence that is statistically more likely to come up first. What's more, he found a quick and easy way for you to identify that sequence. Let's take the sequence your friend picked above, heads, heads, tails. In order to work out which sequence you need to pick to give you the best chance of winning you need to follow these easy steps.

1. Take their sequence, copy the middle value in your head and add it to the front. HHT = HHHT
2. Flip the value of the first item in the sequence, so heads becomes tails and vice versa. HHHT = THHT
3. Finally get rid of the last item in the sequence. THHT = THH

This new sequence of tails, heads, heads is statistically more likely to come up before the sequence heads, heads, tails, with odds of 3 to 1. If you are lucky enough for your opponent to pick head, heads, heads or tails, tails, tails then the odds become 7 to 1 in your favour. And if, for some reason, you don't have a coin it works just as well with playing cards, using red and black in place of heads and tails. Good luck, and enjoy your free drink.

Building your Skeptical Toolkit

What are the chances of that?
The Law of Large Numbers and how we suck at maths.

Now this is going to seem like something of a change of direction after the topics of the last few chapters. However, it does indeed relate to the way we think and our understanding, or probably more accurately our misunderstanding, of the way our minds work. Have you ever heard the phone ring and known exactly who it was on the other end before you picked up the line? Or maybe you have been thinking about someone only to find out later that they died at that exact moment. Could it be that you possess some kind of latent psychic ability? Maybe you are linked to some force, unknown to modern science, that allows you to detect the thoughts of others on some subconscious level, not enough to hear them but enough to know when they are thinking about you or when those thoughts get removed from the general world consciousness. Or maybe, just maybe, something else is going on here.

Let's start with a simple question. What are the odds of you, buying a single ticket, getting all six numbers in the UK National Lottery? I won't bother with the maths[*] and instead will just tell

---

[*] Oh ok I will. It is very simple to work out. There are 49 balls in the lottery draw. The odds of getting one of your six numbers is therefore 1 in 49/6. The odds of then drawing one of your remaining five numbers is 1 in 48/5. The odds continue like this with the number of balls remaining and the number of guesses remaining decreasing as you go along. To work out the odds of getting all six balls right you simply multiple these individual odds together. $(49/6)*(48/5)*(47/4)*(46/3)*(45/2)*(44/1) =$

you. Your odds of winning with just a single ticket are 1 in 13,983,816 or approximately 14 million to 1 against. Just to put that into perspective according to the National Weather Service the odds of getting struck by lightning in any given year are 1 in 750,000. That means you are 18 times more likely to get hit by lightning than you are of getting all six numbers in the lottery. Them are some less than favourable odds. So let's take it a step further and, sticking with the idea that the only way to win is to get all six numbers, ask what the odds are of *anyone* hitting the jackpot?

Well according to the Saturday lottery ticket sales figures for the first three months of 2010 an average of 32,773,630 people played the Saturday lottery every week for the first quarter of that year. Now we know the odds of a specific person winning are 1 in a little under 14 million so all we need to do is divide the number of people playing by the odds of a single person winning. This time things work out a lot better for us with odds of over 2 to 1 in favour of someone getting all six numbers. This means that for any given lottery draw two people are pretty much guaranteed to hit the jackpot.

But what has all this to do with skepticism? Well the point of the lottery example is to show that the odds governing an event can be very different depending on the point of view you are coming from and the questions you ask. The odds of a specific outcome happening may indeed be incredibly high, but the odds of a more generalised outcome can be pretty much a sure thing. The other point of the example is to show that most of us are terrible at calculating these sorts of odds. After all close to 33 million people

---

13,983,816. So your odds of getting all six numbers in the lottery with a single ticket are 1 in 13,983,816, or roughly 1 in 14 million. Still want to buy a ticket?

Building your Skeptical Toolkit

play the lottery each week even though the numbers show that only two of them are likely to hit all six numbers.

Bringing things back to one of the examples at the beginning of the chapter let's apply the same sort on analysis to a question of a more supernatural nature. Almost every year there are stories in the media about so called psychics predicting the exact time of someone's death. You see them in newspapers, though usually not the good ones, and on TV shows presented by people who are recognisable just by their first name. Much is made of these events by those who believe in psychic powers and it is often claimed that there is no way to explain them other than some sort of psychic ability. Well let's throw some maths at the problem and see how it works out.*

Let's say that each of us knows of 10 people who die each year. Now while some of these people will be friends and relatives it is likely that the majority of them will be celebrities or other people with whom we have only a passing acquaintance. Now let's say that you only actually think of each of these people once a year, I mean you are far too busy to think of great aunt Agnes more often than that after all. Ok maths time.

In each year there are 105,120 five-minute intervals. This means that the odds of you thinking of one of these ten people during the exact five-minute interval during which they die is 1 in 105,120. But of course we have ten people to play with here so we have to multiply this number by ten, giving us odds of 1 in 10,512 of you thinking of any one of these ten people during the exact five-minute interval during which they die.

Now these are still fairly long odds. I doubt many of us, National Lottery players excluded of course, would take part in a game

---

* These equations are based upon the work of physicists Georges Charpak and Henri Broch, authors of the book *Debunked!: ESP, Telekinesis, and Other Pseudoscience.*

where the odds of winning were 10,512 to 1 against. But again we have to look at this in the wider scheme of things.

The estimated population of the UK as of July 2009 was 61,126,832. Of this number it was estimated that 50,893,318 were over the age of 14 and so this is the number we will use for the rest of the equation*. We know the odds of you specifically thinking of one of your ten people during the exact five-minute interval during which they die are 1 in 10,512, but what are the odds of any one of these 50 million people doing this? Well again we simply take the odds for one person and divide them by the number of people involved. This works out as 4841 people thinking of someone during the exact five-minute period during which they die *every year*.

This equates to 13 people every single day who are thinking of someone at the moment they die. That some of these 4841 people contact the media to tell of this amazing occurrence says more about their ability with maths than it does their psychic powers. And of course those psychics who were thinking of someone who didn't die don't go to the press so we really don't know how many wrong guesses there were for every right one.

What we have been looking at here is something known as the Law of Large Numbers. The law says, and I am paraphrasing somewhat here, that given enough time and enough chances even the most unlikely of outcomes is likely to happen. Take the United States for example. With a projected population, as of June 2012, of around 313,826,882 it can be logically argued that a one in a million miracle event will happen to at least 313 people every day. But of course we simply don't think of it this way.

---

* The official figures from the 2011 census will not be released until September 2012. As of July 2010 the estimated population of the UK had risen to over 62 million. As such the numbers I'm using above are a little out of date, but I think the point still stands.

Building your Skeptical Toolkit

The law of large numbers is always worth keeping in mind whenever someone tells you that the odds of something happening naturally are simply too high and therefore a given phenomenon must have a supernatural cause. They may indeed be right when saying that the odds of a specific event happening to you at a specific time are incredibly high, but the odds of a similar event happening to one of the seven billion or so people on the planet at some point in recorded history are a lot smaller, maybe even guaranteed.

Another great example of how bad most of us are when it comes to calculating probabilities is found in the Birthday Problem. The birthday problem asks how many randomly chosen people you would need to have in a room before two of them shared a birthday. Stop reading and think about it for a second, what number do you come up with?

Well the actual answer comes as a surprise to most people. In order for the odds of having two people who share a birthday to be over fifty percent you only actually need twenty-three people in the room. Most people find this counter intuitive and will argue against this answer, however the maths is sound. The reason most people think that twenty-three people is too low a number is because they are, once again, looking for a specific answer rather than a more general one. They make the mistake of trying to calculate the odds of one specific person sharing a birthday with another specific person. It's true that the odds of some random person sharing the same birthday as you is 1 in 365, but the odds of any person in a group sharing a birthday with any other person in the group only requires twenty-three people before it is over 50%.[*] Once you get up to fifty-seven people in

---

[*] Again this works in a very similar way to how we calculated the lottery odds. If you just have two people then the odds of them sharing a

your group the odds of two of them sharing a birthday increases to 99%.

However, it is not always bad maths skills that make us susceptible to miscalculating the odds or becoming convinced the chances of something happening are longer than they actually are. Sometimes the information we are presented with has been intentionally twisted to fit a particular bias and if we don't look at the information skeptically we can end up believing all manner of things that might not be completely true, or at the very least not as bad as we have been led to believe. In his book *Bad Science* and his Guardian column of the same name Dr Ben Goldacre has investigated many of the ways in which the media and proponents of pseudoscience can twist the data to support their personal agendas and for the rest of this chapter I want to quickly go over his findings, as well as a more recent example of this in action.

First let me present you with two entirely fictitious newspaper headlines reporting on the same story:

Eating donuts increases your chances of getting eyebrow cancer by 50%

Eating donuts increases your chances of getting eyebrow cancer by 1%

Two questions for you. Firstly how can both of those headlines be reporting on the same story and both be accurate? And secondly which one do you think will sell more papers?

The answer to the second question should be fairly obvious. The media simply loves big flashy headlines, especially ones that

---

birthday are 1 in 365/1. With three people it becomes 1 in 365/2 or 1 in 182. Four people 1 in 365/3 or 1 in 122 and so on.

promote what Ben Goldacre likes to call "miracle cures and hidden scares." Telling someone that something they do will drastically increase their chances of contracting some terrible disease or ailment sells papers much more effectively than telling them the statistical risks of said action are marginal at best. Equally twisting the statistics to over hype the benefits of eating, let's say, broccoli, a supposed "superfood", with every meal is just as likely to equate to increased sales figures. The answer to the first question may not be as obvious but it is the reason why the media can get away with flashy attention grabbing headlines that turn molehills into mountains without actually crossing the line into outright dishonesty.

Let's stick with our entirely fictional donut headlines. In this example the increased risk of getting eyebrow cancer as a result of eating donuts is described as being 50%, which sounds terrible, but also as 1%, which doesn't sound like much of a risk at all. So how can both of these be accurate?

Well let's say that in a group of one hundred people it is likely that two of them will get eyebrow cancer just as a normal run of the mill side effect of life. We then discover that in a group of one hundred people that eat donuts the number of people who get eyebrow cancer is three out of a hundred. In other words eating donuts will cause one extra person out of a hundred to develop eyebrow cancer. Describing it this way it doesn't sound too bad and it is also easy to understand, as it doesn't involve percentages or probabilities, it is just simple numbers that should be clear to anyone. This is what is known as 'natural frequencies' and it is a way of explaining statistical differences in a way that is natural to us and so simple to follow. But there are other ways of explaining the same information.

I could have chosen to describe the same data by saying that the risk of getting eyebrow cancer naturally is 2% and that eating donuts increases this risk by 1%. I don't need to explain, but I will anyway, that 2 people out of a hundred represents 2% of the entire group and that three people, or 3% of the group, is an increase of 1% over the norm. This way of describing the increase in risk is known as the 'absolute risk increase'. Again this doesn't sound too bad but there is still another option I could use.

If we focus on just those two people who are naturally going to develop eyebrow cancer as being 100% of the current risk level and eating donuts increases this to three people, well then that is half as many people again as would have got it naturally or in other words 50% more. This is known as the 'relative risk increase' as it describes the increase in risk relative to the current risk level. You can see why this is the option that the media like to use.

It is important to keep these different ways of reporting the same information in mind when reading those attention-grabbing headlines. The media almost always use the relative risk increase method in their stories and almost never stick with the far easier natural frequencies approach when trying to convince us that something will kill or cure us. To use one of Goldacre's own examples the media delighted in telling us that taking ibuprofen would increase our chances of "heart failure" by 24%. Wow, that sounds scary. I take ibuprofen for both headaches and back pain. Am I putting my life at risk here? Well no, because once again this is the relative risk increase. Only a couple of papers chose to report the story using natural frequencies that explained that this was actually just talking about 1 extra instance of "heart attack" in a group of 1,005 people taking ibuprofen over and above the natural risk of heart attack in a group this size.

Of course there are other reasons, other than selling newspapers, why people might twist the data to give a figure that they find more favourable. Often people have ideological reasons for

Building your Skeptical Toolkit

wanting to word the data in a certain way and this, combined with our general misunderstanding of numbers, allows them to promote their personal pseudoscientific views and sometimes even anti-science misinformation. Let me end this chapter with a recent real world example of this.[*]

In February 2010 CNN ran a story about how over a thousand young people, with an average age of 14 to 18, in the New York, New Jersey area had contracted mumps since August the previous year. The part of the story that we are going to look at here revolves around the fact that some of the people who contracted the virus had received the full course of vaccinations against it.

> The mumps outbreak also spread to Ocean County, New Jersey, where 159 confirmed cases have been diagnosed since September, county spokeswoman Leslie Terjesen told CNN. An additional 70 others are suspected of having mumps, she said.
>
> [...]
>
> Anyone fully vaccinated from mumps receives two doses of the vaccine, according to the CDC. Of the New Jersey cases, 77 percent were vaccinated, Terjesen said.

This part of the story was picked up by Mike Adams, the self-proclaimed "Health Ranger" and editor of NaturalNews.com, who reported on it online and in his email newsletter. It is also worth noting that Mike Adams is also a somewhat prominent member of

---

[*] Credit for this section really needs to go to Skeptico, anonymous author of the blog of the same name, who did all the analysis on this and graciously allowed me to use it in this book. Check out his/her work at Skeptico.com

Your Brain Hates You!

the anti-vaccination movement that has seen a resurgence in popularity in recently years, an upturn that interestingly correlates with the increase in preventable virus outbreaks. I think his personal views on the subject are clear in his treatment of this story (emphasis original).

> To hear the vaccine pushers say it, all the recent outbreaks of mumps and measles are caused by too few people seeking out vaccinations. It's all those "non-vaccinated people" who are a danger to society, they say, because they can spread disease.
>
> Reality tells a different story, however: It is the vaccinated people who are causing these outbreaks and spreading disease!
>
> Just this week, an outbreak of mumps among more than 1,000 people in New Jersey and New York has raised alarm among infectious disease authorities. The outbreak itself is not unusual, though. What's unusual is that the health authorities slipped up and admitted that *most of the people infected with mumps had already been vaccinated against mumps.*
>
> In Ocean County, New Jersey, county spokeswoman Leslie Terjesen told CNN that **77 percent of those who caught mumps had already been vaccinated against mumps.**

To clarify these comments, and those made in the rest of the article, Adams is saying that because 77% of the kids that contracted mumps in this outbreak had been vaccinated you are

Building your Skeptical Toolkit

therefore more likely to get the mumps if you are vaccinated than if you are not. Adams puts it bluntly:

> "Vaccines may actually increase your risk of disease. Notice that far more vaccinated children were stricken with mumps than non-vaccinated children?"

Could he be right? Could vaccines actually make it more likely that you will contract something like mumps? Short answer? No. Longer answer? Hell No. Useful answer? Well this requires us to apply the lessons we have learned about how statistics can be reported in numerous different ways, some more useful than others, and for us to use just a little bit more maths.

For a start Adams has come to the wrong conclusion, partly because of his pre-existing bias regarding the subject and partly because, as we learnt earlier in the book, he is asking the wrong question. Adams is asking *"did more kids who had been vaccinated get mumps than kids who had not been vaccinated?"* Well the answer to this question is yes. 77% of the kids infected had been vaccinated while only 23% of the infected had not been vaccinated. At first glance this seems to prove his point. However the question he should have asked was *"are you more likely to get mumps if you are vaccinated?"* This question provides a very different, if slightly more complex, answer. Let's get the maths out.

Firstly let's put those percentages into numbers that are easier to understand. Now unlike Adams implies, that number of 77% only actually applies to the 159 cases in Ocean County, New Jersey. As such 77% of 159 works out as 122 children who caught the mumps and were vaccinated while only 37 of those who got it were not vaccinated. Things still seem to be going in Adams favour but let's move on.

271

Your Brain Hates You!

According to CDC reports, that are freely available online, 96.1% of New Jersey kindergarten pupils are fully vaccinated against the mumps. Let's round this down to 96% to make things a bit easier for us all. Using this number we can work out that this means that the number of vaccinated children in New Jersey is 24 times, that's 96 divided by the remaining 4 percent, the number of unvaccinated children. This is a very important number to keep in mind.

Imagine that we have a group of 1,270 children who are exposed to the mumps.* Of this number we know that on average 96% will be vaccinated. This gives us 1219 children that have been vaccinated and 51 children, the remaining 4%, which are non-vaccinated. In New Jersey we had 159 children who caught the mumps of which 77% or 122 children had been vaccinated. So how does this help us answer our question? Well it breaks down like this:

Children vaccinated and who got the mumps
122 / 1219 = 10%

Children unvaccinated and who got the mumps
37 / 51 = 73%

As you can see when asking the right question, that being *"are you more likely to get mumps if you are vaccinated?"* it becomes clear that the answer is an emphatic no. In fact you are 7 times as

---

* This number is not just picked at random and I think needs a little more explanation. According to the CDC the effectiveness of MMR against mumps is approximately 90% after two doses. As the vaccination rate is 96% starting with 1270 children this works out as 1219 vaccinated children of which for 10% the vaccine will not be effective. 10% of 1219 is approximately 122 children, the number of vaccinated children infected in New Jersey.

Building your Skeptical Toolkit

likely to get mumps if you are unvaccinated than you are if you are fully vaccinated against mumps. So how do we explain the 77% figure quoted by both Adams and CNN? Simple, there were more vaccinated children exposed to the mumps than non-vaccinated children, 24 times as many. If Adams were right and the vaccine against mumps actually makes you more likely to catch the virus then you would expect the number of children both vaccinated and infected to be 24 times as large as it is. As it works out the number of vaccinated children with the mumps is only three times, 77 divided by 23, the number of non-vaccinated children with the virus.

It is also worth pointing out the fairly obvious fact that the higher the percentage of children that are vaccinated is then the higher the percentage of kids who get mumps and who are also vaccinated will be. If 100% of children are vaccinated against mumps then it stands to reason that 100% of the children that catch the virus will also be vaccinated. This doesn't mean the vaccine doesn't work the vast majority of the time or that it actually makes you more likely to get the virus, it simply means that with an imperfect vaccine, being only 90% effective, the more kids you have that are vaccinated then the higher the chance is that one of them will catch the virus they are vaccinated against.

When it comes to numbers reported in the media or someone quoting odds at you it always pays to take a moment and be skeptical. Are the odds they are calculating accurate? Are they taking into account all of the possible outcomes or just focusing on one specifically? Do the numbers they are presenting to you represent the relative or the absolute level of risk? Is there a reason they are not simply telling you the natural frequency numbers? Keep these things in mind and it should be fairly easy to discover the truth behind the numbers.

Your Brain Hates You!

Oh and vaccinate your children.
Seriously.

Building your Skeptical Toolkit

Skeptical Sidebar #6
The Monty Hall problem

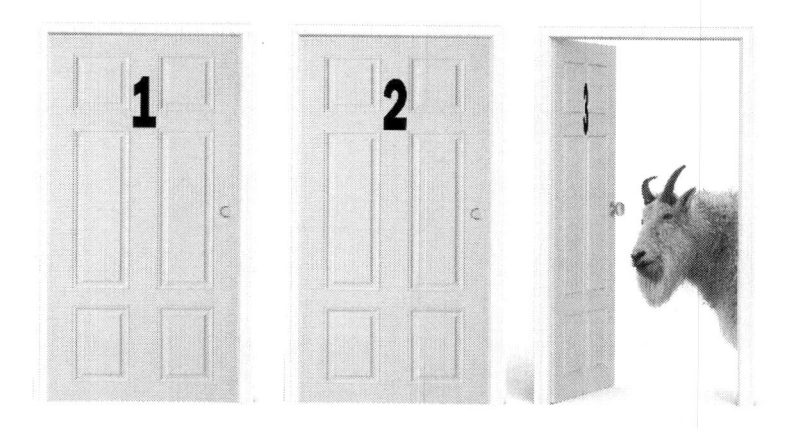

So as we have seen when it comes to calculating odds most of us don't do a very good job at it. As such let's look at a much simpler odds related question and see if we can improve your chances of getting it right, and in this case of winning a car as well. We're going to try a little puzzle known as the Monty Hall Problem. The puzzle is named after, surprise, surprise, Monty Hall who used to host the TV quiz show "Let's Make A Deal". Now technically how this puzzle works is not the same as how Monty did things on TV, but the general idea is similar enough.

The puzzle goes like this. Monty presents you with three closed doors and informs you that behind one of them is a fabulous sports car or luxury holiday or whatever prize takes your fancy. However behind the other two doors is a goat. All you have to do is pick the door with the prize behind it and you win the car/holiday/whatever. You make your selection, let's say you choose door number 1, but don't open it at this time. Monty then opens one of the remaining two doors; let's go with door number 3, and reveals that there is a goat behind this door. Now he asks you the all-important question. Do you want to stick with your original choice of door number 1 or do you want to switch to door number 2?

The aim of the puzzle is to work out what action gives you the best odds of winning. Should you stick with your original choice? Should you change to door number 2? Or does it not matter either way? Take a moment to think about it and then we will have a look at the answer.

Stick or Switch?

Your Choice   Monty's Choice

Now the chances are that you, like the vast majority of people who attempt this problem including many PhDs, came to the conclusion that it really doesn't matter what you do as your odds of getting the prize are the same either way. You know it is not behind door number 3 so that leaves just two places it could be. This means that there is a 50/50 chance that it is behind door number 1 and a 50/50 chance of it being behind door number 2. So the odds are identical whether you stick or switch. If this is the answer you selected then congratulations, you are completely...wrong.

Despite this being the intuitive answer it is in fact incorrect, you are actually twice as likely to win if you switch to door number 2 than you are if you stick with door number 1. So there you go, that is the Monty Hall problem solved, you should always switch. What? Oh, you want me to back up my answer with evidence huh? Good for you, it seems we will make a skeptic of you yet.

Ok so here is how it works. When you are first presented with the three doors each door has a one in three chance of being the door with the prize behind it. At this point your odds of picking the right door are the same no matter which door you select, and we would expect you to pick correctly once out of every three attempts.

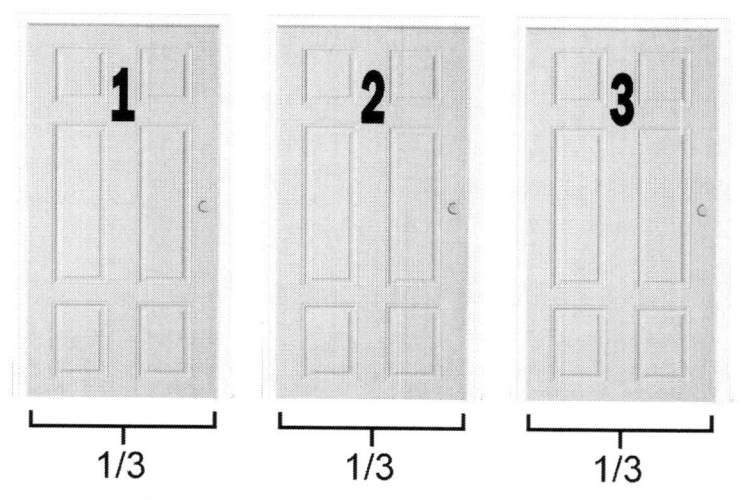

1	2	3
1/3	1/3	1/3

Now comes the time to pick a door and we will once again pick door number 1. As already mentioned your chance of picking the right door first time is 1/3, which means that your chance of not picking the right door first time is 2/3.

Your Brain Hates You!

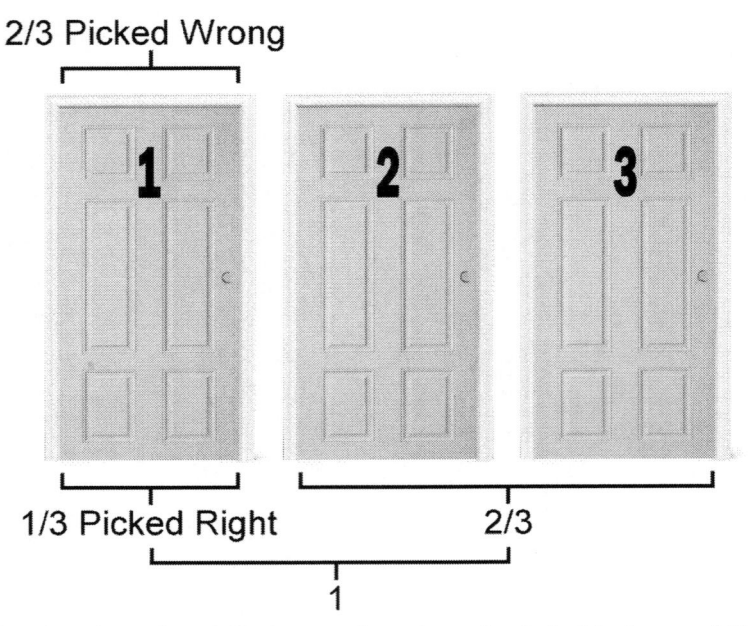

2/3 Picked Wrong

1/3 Picked Right   2/3

1

Likewise there is a 2/3 chance that the prize is behind one of the other two doors. As such we can break things down as in the image above. Now I am sure at this point you will all agree that it is twice as likely that the prize is behind one of the doors you did not pick than it is behind the door you did pick. For most people the problem comes when Monty steps up and opens one of the doors. Once again we will have Monty open door number 3 to reveal the goat waiting behind. This means that the prize must therefore be behind either door number 1, the one you picked, or door number 2. The mistake most people make is to assume that it is therefore now a two horse race and that the odds can be divided up accordingly between the two doors, 50/50. But this is not the case and in fact the odds of you having picked the correct door remain exactly as they were before Monty opened his door. Here's how it looks now.

Building your Skeptical Toolkit

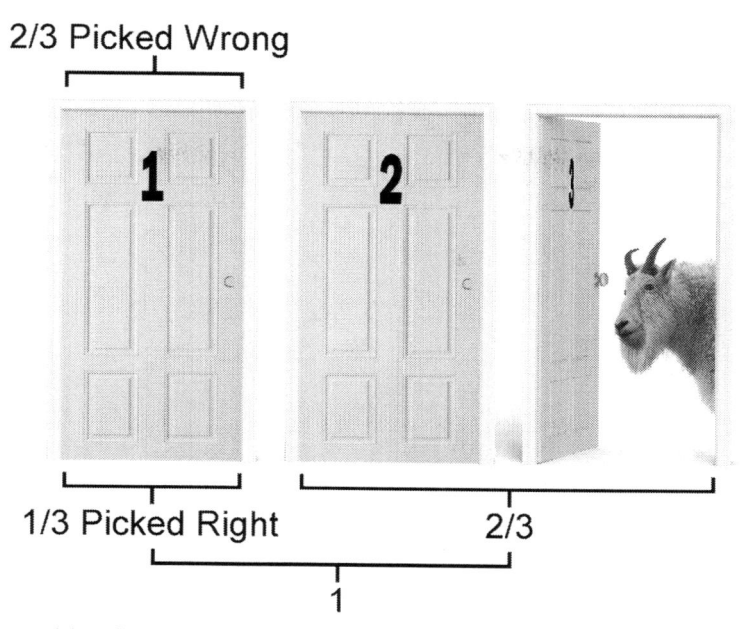

2/3 Picked Wrong

[Doors 1 and 2 closed; Door 3 open with a goat]

1/3 Picked Right   **2/3**

1

The odds of you having picked the correct door the first time are still 1/3 and the odds of you having picked the wrong door, even though you now know that the prize is not behind door number 3, are still 2/3. And because we do know that it is not behind door number 3 then the odds that the prize is behind door number 2 become 2/3, twice the odds of door number 1. Still not convinced? Well ok let me put it another way. Here is a table showing the various different outcomes for the three possible arrangements of the goats and prizes, given that you start by picking door number 1 each time and that Monty always opens a door with a goat behind it.

Door 1	Door 2	Door 3	Result if you switch	Result if you stick
Car	Goat	Goat	Goat	Car
Goat	Car	Goat	Car	Goat
Goat	Goat	Car	Car	Goat

279

Your Brain Hates You!

As you can see you are again twice as likely to get the car if you switch rather than sticking with your initial choice. And there you go, once more you can see just how bad most of us are when it comes to correctly calculating the odds of, well, pretty much anything really.

Building your Skeptical Toolkit

Let me tell you your future
The less than magical art of cold reading

I'm going to try something a bit different now; something that I don't think has ever been tried before. Before I tell you what I'm going to attempt maybe I should give a little bit of an introduction. I am sure you have all heard at some point in your lives that there exists a special bond between an author and his or her readers. When writing a book an author cannot help but give away information about themselves, information that the reader often only picks up on a subconscious level.

By this point you probably have a fairly good idea about what type of person I am. You could probably describe fairly accurately the kind of things that I am interested in, what I value and deem important as well as those things that get me fired up and annoyed. You may even have an idea about my sense of humour as well as where I stand with regards to political and religious issues. You probably know all of these things without even realising it and if I were to ask you to write down a description of the kind of guy you think I am I'm pretty sure that most of you would come up with something that was fairly spot on.

Well what you may not be aware of is that this bond works both ways, enabling me to know information about you simply by way of you reading this book. And that's what I am going to try to do now, use that bond to describe you, the person reading this book right now. Ok, I can tell that you're skeptical, or at least by this point in the book you damn well should be, but I ask that you open yourself up to the possibility that I am actually able to do what I say I can do. Remember a good skeptic has an open mind and is willing to accept that there are things that they do not know and that things they believe in may in fact be wrong. So take a deep breath, make sure that you are comfortable and open up your mind. Be aware that the connection I have with you will vary from reader to reader, as some of you will be able to open yourselves up to the process more than others, and the things I see are not always clear and as such may not make sense to you at first, but I ask that you think about them and see if they do fit with aspects of your life. Anyway let's get on with it shall we, let me read your mind.

> I sense through the book that you are generally a happy and easy going person, someone people like once they get to know you, though you don't always make that process as easy as you could. I see you as someone who has a few close friends rather than a large number of casual acquaintances, but this is not a bad thing. I sense you are a better friend than you sometimes have reason to be and that your friendships are something you will passionately defend should the need arise.
> I think this has much to do with the fact that you are generally a bit more honest and conscientious than most people. Don't get me wrong I'm not saying that you're a saint and you know there are times when you can be a little bit selfish. It is just that when it really matters you are someone who really understands the importance of being trustworthy. For the most part you have a cheerful and positive attitude, though if you are honest with yourself there have been times when this has been far from the case. Generally you let most things slide off your shoulders and it tends to take a lot to get you actually angry. That said if someone challenges one of

Building your Skeptical Toolkit

your core beliefs or values they'd better watch out as you have quite a temper once you get going.

I sense that you are a wise person, someone who is more shrewd or perceptive than many of those around you, and definitely more so than you are given credit for. However I don't feel that this wisdom comes from what you might call "book learning", but more that you are wise in the ways of the World, something that doesn't apply to everyone. I also feel that you are intelligent enough to recognise that wisdom comes to us in many ways and in many different forms and that it is not all about formal education.

I get the feeling that you are good at what you do and that this is recognised by your superiors, though not always in what you would consider a positive way. I sense that you work harder and likely have a better understanding of thing than some of those around you. You are not someone who simply skates by doing as little as possible and you sometimes find yourself frustrated by those who do. Saying that you sometimes wonder if this really is the right role for you. You wonder if you could be doing something else with your life, something that would perhaps challenge you more than your current role or that would be more in line with your personal interests and values.

Looking back to your childhood I am seeing an accident, or maybe an illness, nothing overly serious but something that had your parents far more worried than they had need to be. But thankfully this is not something that had a lasting effect on you and I sense that this is now little more than a distant memory. I also feel that when you were younger there was something in which you had a lot of interest and that you had a certain amount of talent for. Unfortunately as you have grown older the pressures of life have forced you to put this to one side and it never really grew into the great thing it might have been.

Turning to your home I can see that you have a collection of photos, the vast majority of which remain unsorted and undisplayed. I sense that you also have a couple of gadgets, electronic gizmos or the like, lying around the house that don't work and that, if you are honest, you will never get around to having fixed. Finally in relation to your house I also see a blue car parked outside. I'm sensing some connection to this car, though perhaps not directly, but I think it being outside your house will make sense to you if you think about it.

Let me end by making a prediction. A week after finishing this book you will bump into someone who you haven't seen in a while who will tell you about a big change that is about to happen in their life that relates to their job.

Oh ok, so obviously I wasn't actually sensing you through the pages of this book, or was I? Either way I feel pretty sure that for many of you my description sounded eerily accurate. So how could that be seeing that, hopefully anyway, I don't know everyone who is reading this right now? Well what I have actually done here is utilise a couple of the elements that makes up a skill used by so called psychics and mediums all over the World. Cold Reading.[*]

---

[*] Much of the information in this chapter is taken from an excellent book on cold reading entitled *The Full Facts Book of Cold Reading* by Ian Rowland. In his book Ian covers 38 different elements that make up cold reading. I am only going to cover a few of them here. Also most of the terminology in this chapter comes directly from Ian's book and as such it is possible I may discuss concepts that you know of under another name. But don't worry about this, as it is the techniques themselves that matter and not what they are called.

Building your Skeptical Toolkit

Now technically cold reading doesn't really come under the heading of ways that your brain fools itself, in fact it is probably more accurately described as other peoples brains fooling you. However, the reason that cold reading works has a lot to do with the things we have looked at in the previous chapters and as such I think it is worth spending quite a few pages looking at this tricky little subject. Put very simply cold reading is a series of psychological techniques that, amongst other things, allows you to create the illusion that you know specific information about someone that they believe you couldn't possibly know by normal means. In fact a number of the different techniques that make up cold reading prey on the various mental weaknesses we have looked at in the previous chapters. Cold reading is generally used by psychics[*] as a way of convincing people that they really do have special powers and really are talking to your departed loved ones. Many cold readers may also choose to make use of a number of different props in order to reinforce the idea that they are connecting with some otherworldly power. These props include things like tarot cards, crystal balls or astrological charts, but the important thing to remember is that these are props and are simply ways of framing the actual skill at work, that of cold reading.

I should also quickly mention some of the things that cold reading is not, as a lot of people have a number of misconceptions when it comes to this topic. Cold reading is not about reading body language or Sherlock Holmes style eagle eyed observations. While these things may play a small role in the process this is not

---

[*] Something else I am going to borrow from Ian's book is the way he handles mention of psychics. Rather than add "so called" or "alleged" in front of the word every time I use it I am simply going to refer to people who use cold reading techniques as psychics. Also, simply for the sake of consistency and nothing else, I will always refer to both the psychic and the person getting the reading using female pronouns.

primarily what cold reading is all about, no matter what TV has told you. Equally cold reading is not about fishing for information or making vague generalisations that could apply to anyone, though again both of these things do play their part. In fact done well cold reading can produce what appear to be incredibly specific and insightful information. People who are taken in by cold readers are not fools, far from it. As I have said a number of times in this book often the only thing that stops people from falling victim to this sort of thing is an understanding of what is really going on and a healthy dose of skepticism. Intelligent people can be taken in just as easily as anyone else. Take someone like mentalist and performer Derren Brown for example. He uses cold reading techniques in his shows and actually informs people of this fact, making it clear that everything he does is a trick, beforehand. And yet without fail he proceeds to amaze people with scarily accurate details about their lives that he really shouldn't know to a level that results in many people leaving his shows convinced that he really does have psychic powers and that the claim that he is using cold reading is in fact the deception. Cold reading is a skill, and an impressive one at that. If you don't know what is going on then have no doubt that you will be deceived.

One more thing that is worth mentioning before we move on is the idea of hits and misses. Numerous times in this book when mentioning psychics I have talked about the way that people tend to remember the hits, those statements made by the psychic that appear correct, and forget the misses, those statements made by the psychic that are incorrect. I want to make it clear here that this is something that the person getting the reading does and not something the psychic themselves does. Psychics do not count hits and misses; instead they treat everything as a hit and have numerous ways of turning even the most obviously incorrect statement to their advantage. But I am getting ahead of myself and we will look at what expert cold reader Ian Rowland calls the "Win Win Game" later in the chapter.

So let's start by looking at the techniques I used in my psychic reading at the start of the chapter. Limited by the one way interaction of the written word my reading is a lot more generalised than the average cold reading session is likely to be, but nonetheless it forms a good starting point from which to look as some of the cold reading techniques you may already be familiar with.

It may surprise you but the first techniques I've employed are not found in the reading itself, but rather in the introduction to it. A good set-up is an incredibly important part of the cold reading process and can mean the difference between a successful reading and complete failure. The set-up to a cold reading session is primarily about getting the subject to relax and encouraging them to cooperate with the proceedings. The psychic will do their best to be likeable and establish a friendly "I like you, you like me" connection right off the bat. Many psychics have specially prepared rooms with low lighting, soft music and comfortable chairs, all of which is designed to relax the subject and convince then that there is nothing to worry about here, that it is just two friends sitting down for a chat. This is not un-similar to what many psychiatrists do. By presenting a friendly relaxing environment it makes the subject or patient much more open to what the psychic or psychiatrist has to say.

The psychic will then run through a series of steps designed to encourage the idea that what is going to happen is genuine and that the psychic really can do the things they claim. The first of these, and the one I missed from my introduction, is for the psychic to establish her own credentials. This usually involves letting the subject know that they have been doing this for a number of years and have had many happy clients. They may mention organisations of which they are a member or have certificates and books displayed that convey a sense of legitimacy to the proceedings. High quality props, such as expensive looking tarot cards, can help sell the idea that you are dealing with a professional, just as an expensive guitar, for example, implies that the person is a serious musician rather an a weekend hobbyist.

Once the psychic has sold herself to the subject it is time to sell them on the specific belief system they intend to use in the reading. This may simply involve a quick history of the technique, how it has been used for hundreds of years and is viewed as one of the most effective methods for whatever it is the psychic wishes to claim it is used for. In my introduction I discussed the bond that exists between author and reader, something which I believe there is good reason to expect, and then expanded on this idea to explain how I would do the reading, thus again giving my claim of psychic powers a sense of legitimacy.

Next I asked for you to open yourself up to the proceeding and cooperate with me in analysing the things I had to say, encouraging you to seek meaning in my words even when the things I said did not at first seem relevant to you. And this is exactly what a real psychic would do. A common approach is to say that the things the psychic sees are not always clear and will likely mean more to the subject than to the psychic herself. As such, the psychic claims, she will require the subjects help in correctly interpreting the meaning behind what she comes out with. This puts pressure on the subject to make sense of the psychic's words and many people will come up with things that barely relate to what the psychic said as a way of being helpful and alleviating the discomfort of the psychic being wrong.

The final thing I did in my introduction was make excuses as to why it might not work. This way I had a pre-existing out should my reading prove to be inaccurate. Many psychics will come straight out and admit that sometimes things go wrong and that they are not claiming that they will get it right every time. They may also, as I did, imply that any failings in the reading are partly the fault of the subject herself. For example I said that my reading would only work if you really opened yourself up to the process. Thus if my reading was incorrect then the fault lies with you not being open enough and not with any flaw on my side of the equation.

Once this is done and the subject is suitably relaxed and ready and willing to provide feedback to the psychic it is time to start the reading itself and this is where the various techniques that make up cold reading really come into play.

Barnum statements

I used a lot of Barnum statements in my reading. Barnum statements, named after American showman P.T. Barnum, are statements that seem personal and yet apply to the vast majority of people. They also tend to appear specific but are designed to be as open as possible. These are your basic high probability hits. Most people, simple due to the fact that we are all human, share some experiences and tendencies in common. Combine this with the subject's willingness to fill in the details, specifically requested during the set-up, and the psychic is practically guaranteed a hit. And even with a miss it still tells the psychic something about you. For example some of the Barnum statements I used were:

> "Looking back to your childhood I am seeing an accident, or maybe an illness."

> "[You] sometimes wonder if this really is the right role for you."

> "I can see that you have a collection of photos, the vast majority of which remain unsorted and un-displayed."

All those things sound initially to be very specific, and yet given more thought you can see that they actually apply to most people to some extent. Many people suffered illness or had an accident as a child and who doesn't wonder if they should be doing something different to what they are doing from time to time? Let me give another example:

> "Your father passed on due to problems in his chest or abdomen."

Again at first this seems very specific, that is until you think about it a bit more. Someone's death being related to the chest or abdomen covers a wide variety of possible causes such as heart disease, pneumonia, diabetes and most forms of cancer. Additionally even if the subject disagrees and says that her father died of a stroke the psychic can simply turn it around by saying something like "Yes, but the stroke was caused by a problem with his heart." As I said before a practiced psychic can turn almost any response from the subject into a hit.

The Rainbow Ruse

The rainbow ruse is where the psychic makes a statement that appears to award the subject with a specific personality trait but also includes the opposite of that trait at the same time. For example I could say that you are happy person but sometimes you are sad. The name comes from the fact that the psychic can cover a whole spectrum of possibilities in one go. This can make it appear as though the psychic has made an accurate statement about the mind of the subject, even though the actual statement is vague and contradictory. This works because very few people sit at the extremes of any particular personality type or emotion and because, again, most people have experienced both sides of a particular emotion at some time in their lives simply due to the fact that they are human. Examples of this in my reading include:

> "[You] are generally a bit more honest and conscientious than most people. Don't get me wrong I'm not saying that you're a saint and you know there are times when you can be a little bit selfish."

> "For the most part you have a cheerful and positive attitude, though if you are honest with yourself there have been times when this has been far from the case."

> "Generally you let most things slide off your shoulders and it tends to take a lot to get you actually angry. That said if someone challenges one of your core beliefs or values they'd better watch out as you have quite a temper once you get going."

So to recap you are honest but can be selfish, happy but can be sad and you don't get angry but sometimes you do. A psychic can pick almost any personality trait as well as it's opposite and combine them both into a sentence and it will sound specific and yet still apply to the vast majority of people.

Fine Flattery

Fine flattery is basically a way of getting people to agree with what you are saying by simply saying nice things about them. Of course it is a bit subtler that just saying things like "you're great" and often involves comparing them to other people or the general public in some favourable way. Fine flattery also tends to focus on positive aspects of personality that most people tend to think apply to them, such as honesty, being fair-mined, hard-working or loving. In the previous section on the rainbow ruse you'll find an example of fine flattery in the first sentence of the first example. Here I used the positive personality traits of honesty and conscientiousness and stated that you displayed these things more than most people. It is unlikely that any of you felt the urge to correct me and point out that actually you are completely dishonest, and had you done so that would have been rather ironic.

There are two much better examples of fine flattery at work in my reading that deal with the subjects of wisdom and friendship. These are particularly strong examples as pretty much everyone likes to think of themselves as wise and a good friend.

> "I sense you are a better friend than you sometimes have reason to be and that your friendships are something you will passionately defend should the need arise."

> "I sense that you are a wise person, someone who is more shrewd or perceptive than many of those around you, and definitely more so than you are given credit for. However I don't feel that this wisdom comes from what you might call "book learning", but more that you are wise in the ways of the World, something that doesn't apply to everyone. I also feel that you are intelligent enough to recognise that wisdom comes to us in many ways and in many different forms and that it is not all about formal education."

That second example also shows how framing something in the language of the psychic is itself a very effective tool. All I am really saying there is "you're smart" and yet by wording it as I have done it implies a deeper understand of who you are as well as making it less obvious that all I am actually doing is paying you a compliment. Another reason that fine flattery works is due to something we have already looked at, the better-than-average effect. When it comes to almost every aspect of life, be it sporting ability, intelligence or friendliness, most of us are average, after all that is the very definition of the word. And yet when asked to rate ourselves on these things compared to others around 60-85% of us will rate ourselves as being higher than average, when obviously this cannot be the case. Fine flattery simply plays on this natural instinct to see ourselves as better-than-average by telling us outright that this is the case. Very few people can resist a compliment.

The Good Chance Guess

This element of cold reading has more to do with actual facts rather than aspects of the subject's personality. As the name implies these are basically guesses but ones that have a high likelihood of being accurate, at least in some way or another. Here is the example I used in my reading:

> "Finally in relation to your house I also see a blue car parked outside. I'm sensing some connection to this car, though perhaps not directly, but I think it being outside your house will make sense to you if you think about it."

This is a complete guess on my part but one that I feel pretty confident in making. While stating that there is a blue car outside your house sounds like a very specific claim that is either accurate or not you will notice that I have given myself a lot of wiggle room with my choice of wording. I don't say that this blue car is your car, only that you have a connection to it. Now of course this would include the possibility that it is indeed your car, but also that it belongs to another member of your family, one of your friends, a next-door neighbour whose parking space is near your house or even the delivery man who recently dropped off your new TV. All these things could count as a hit.

And of course I've given myself even more room for interpretation by simply stating the colour of the car as "blue". Blue covers such a wide range of shades and tones that it is pretty much a guarantee that someone you know will have a car that you might call "blue". Everything from midnight blue through to aquamarine or a very light shade of cyan could be counted as blue. Some of the more generous amongst you may even have let me get away with various shades of green under the heading of "blue". So yes, technically it was a guess, but not really.

Fuzzy Facts

Ok now we are coming to cold reading techniques that, due to the limited nature of written readings, are not included in my reading but which are none the less worth mentioning. A fuzzy fact is something that appears to be a factual statement and which is worded in a way that makes it likely to be accepted and yet is open enough to cover a number of possibilities, thus enabling the psychic to develop it into something more specific later on. Fuzzy facts can cover a wide range of topics and due to the fallibility of human memories are likely to be remembered as much more specific, direct statements than they actually were. For example the psychic might employ a fuzzy fact that goes something like this:

> "I'm sensing some connection to America or the Americas. Now why might this be?"

Now this is a very general statement as not only does the psychic not specify which part of the Americas it is she is talking about she also doesn't mention what the connection is, that part will be filled in by the subject herself. In fact you will notice that the example ends with a direct request for the subject to fill in the blanks, but we will come onto that in a moment. The important part here is to mention a distant part of the World from where the reading is taking place and to which the subject may have some connection. Let's continue the reading and see how the psychic can build this fuzzy fact into something very specific.

> "I'm sensing some connection to America or the Americas. Now why might this be?"
> "Could it be Canada?"
> "Well the connection I'm feeling does have a distinct northern feel to it but I wasn't too sure about that. I seem to be picking up on Toronto for some reason."
> "Well I am going on holiday to Canada, but not to Toronto. I'm going to Calgary."
> "But it is a part of Canada you would like to visit, maybe that is why I was picking up on that. But Calgary makes more sense with what I'm getting. I'm sensing that you are going there for a specific reason."
> "Yes, I want to go to the Calgary Stampede."
> "Ah yes, that would explain all the cowboy hats."

The psychic starts with something very vague and nebulous and turns it into something very specific.

The Push Statement

While there are many different ways in which the psychic can appear to know specific facts and details about a person's life I wanted to include this one because it works in what at first glance appears to be a counter intuitive way. While most of the other methods revolve around the psychic starting with a general piece of information that could apply to many people, such as common childhood experiences, folk wisdom, cultural trends or seasonal aspects of life, and then tailoring them to fit the subject, with push statements the aim is actually to get a negative response, at least at first.

As such push statements can be very specific indeed and include lots of details about an event or place allegedly from the subjects past. It doesn't matter if the subject rejects what the psychic says as that is not the aim here, instead the psychic is trying to convince the subject that she knows something about her life that the subject herself has forgotten. The psychic accomplishes this by confidently pushing the idea on the subject while at the same time expanding it until the subject suddenly remembers something that fits. Once the subject finds one aspect of the statement that fits they will tend to focus on this and forget the aspects of the push statement that remain inaccurate, or even conclude that the other parts of the statement are also correct and that they just can't remember them. This can be an incredibly powerful effect. After all it is one thing for a psychic to know things that you know, but to know things that you yourself had forgotten! Now that's special.

Building your Skeptical Toolkit

Because it is rather difficult to come up with an effective push statement it is unlikely that the average psychic will have more than a couple of them in their repertoire, but those that they do use will have been honed by years of practice to the point that they are almost guaranteed to get the intended result. One example is the shoe at a party. This push statement involves a story in which the subject breaks a shoe in some way at a party of some sort and their subsequent feelings towards this event. Now while the subject may at first be unable to remember breaking a shoe at a party if they think hard enough, and the psychic acts confident enough in her statement, they may remember having some sort of shoe problem at a social event of some sort. Suddenly this is what the psychic was talking about all along and the subject is suitably amazed.

The Direct Question

Of course a large part of cold reading is getting the subject to reveal information about themselves without really realising it and then feeding that information back to them as though you came about it by supernatural means. As always there are numerous ways for the psychic to do this, but I'm going to focus on three of the most common. Direct questions are the simplest method for a psychic to use and for me to explain as it pretty much involves just asking for the information you want to know. These are generally used at the start of a reading in order to get an idea of the direction the subject is hoping the reading will take. The psychic will often frame these questions as a way of helping the subject get "value for money" out of the reading. They may do this by saying something along the lines of:

"There are many things we could look at today, however a lot of people come with a specific issue that they would like to address. Now while the spirits generally guide me in the right direction it can take time to get to the issues that people really care about, after all we all have so much going on in our lives. Now I am happy to do this but this is after all your time and money we are talking about here. As such is there any particular topic you would like me ask the spirits to focus on?"

As I said these sorts of direct questions tend to be used only at the start of the reading to help the psychic establish a jumping off point for the proceedings. If the subject is suitably open to the whole thing they can provide the psychic with a wealth of information from which to draw. And if the subject seems particularly susceptible to direct questions then there is nothing to stop the psychic from littering the rest of the reading with them as well.

The Incidental Question

I actually used an incidental question when I was talking about fuzzy facts on the previous page. An incidental question is a question that it tacked onto the end of a sentence or statement in a conversational or, well, incidental sort of way, almost as a throw away comment. The aim is to promote feedback from the subject so that the psychic can tell if they are on the right track or not. The example of this I used was:

"I'm sensing some connection to America or the Americas. Now why might this be?"

Building your Skeptical Toolkit

During the set-up the psychic will have explained that the things they see will often make more sense to the subject than the psychic and as such the subject will be ready for this sort of question. The implication is that the psychic already has the information and that she is just asking for clarification of what it might mean, rather than trying to illicit information, which is actually what she is doing. Other examples might include:

"...can you relate to this?"
"...is this making sense to you?"
"...does this sound correct to you?"

There is a second sort of incidental question that aims to gather specific details rather than just confirm that the psychic is on the right track. Now it may seem obvious what I am doing when it is written down like this, but when presented by a skilled psychic in a neutral, by the way, conversational tone, questions like this can slip by almost unnoticed. Examples might include:

"...how might this be significant?"
"...who might this refer to?"
"...and what in your life might this relate to?"
"...why might I be getting this impression?"

Again the implication is that the psychic already has the information and is just seeking clarification, when in fact it is the subject that is providing everything the psychic needs to know.

The Vanishing Negative

I really like this form of questioning as I think it demonstrates both the deceptive nature of cold reading as well as how it uses clever little aspects of human psychology, such as the fact the most of us have trouble keeping track of negatives in spoken conversation, in order to gain results. Let's jump straight to an example.

> "Ok now let's move onto your job. You don't work in a bank do you?"
> "No I don't."
> "I didn't think so. That doesn't really fit with the feeling I'm getting."

Well that seems fairly straight forward. But what if the psychic gets a different answer?

> "Ok now let's move onto your job. You don't work in a bank do you?"
> "Why yes, yes I do."
> "I thought so. I'm getting a real sense of your involvement in financial matters."

In both cases the psychic uses the exact same question, only in the second example the negative part of the question is simply forgotten. This is a very effective way for the psychic to ask a specific question and yet have all her bases covered at the same time. Now a skilled psychic will expand on this idea with a technique that expert cold reader Ian Rowland calls "re-affirm, re-assure, expand". Let's stick with the same example and see how this can work.

> "Ok now let's move onto your job. You don't work in a bank do you?"
> "No I don't."

Firstly the psychic re-affirms that this is what they were seeing all along.

> "I didn't think so. That doesn't really fit with the feeling I'm getting."

Now the psychic reassures them that this is a good thing, and maybe throws in a little bit of a dig at the alternative as well.

> "It is not an area in which I really feel that you would fit, all those straight laced, by the book people. You strike me as much too interesting for a job like that."

Finally the psychic expands on the original idea in order to imply that they knew the answer to the question all along.

> "Anyway the reason I mentioned the bank is that I can see a positive change in your financial situation as it relates to your job."

Now let's run through the same process again, only this time with the second answer we got to the question.

> "Ok now let's move onto your job. You don't work in a bank do you?"
> "Why yes, yes I do."

Firstly they re-affirm that this is what they meant all along, and in this case drops that pesky negative.

> "I thought so. I'm getting a real sense of your involvement in financial matters."

Next she reassures them that they are doing the right thing.

> "It takes a special sort of person to work in an area like that. You have to have a real head for numbers and that is something not everyone has. Most people just don't have the smarts for your line of work."

And then lastly they expand on the idea as if they knew the answer all along.

> "Anyway the reason I brought up your employment at the bank is that I can see a real upturn in your working life in the near future."

With this simple method the psychic scores a hit no matter what answer they get.

The Win Win Game

Which brings me on nicely to something I mentioned earlier, and that's how psychics don't count hits and misses but instead turn everything into a hit. This is known as the win win game and as always there are numerous different techniques the psychic can use to make sure that they are always right.
Let's say out psychic comes out with something that really doesn't mean anything to the subject. For example let's imagine that the psychic says they are getting the name Brian and yet the subject can't think of anyone of that name. It seems as though the psychic has scored a miss, but by using a technique called "persist,

Building your Skeptical Toolkit

wonder and let it linger" they can turn it around and make it seem as though the fault lies with the subject rather than with their psychic ability. With the Brian example it might go like this.

> "I'm getting the name Brian, does that mean anything to you?"
> *"No, I don't know anyone by that name."*
> "Are you sure, I am definitely getting the name Brian. You are sure you don't know anyone by that name?"
> *"No, I definitely don't know anyone called Brian."*
> "You're completely sure. I'm getting the name Brian quiet clearly. You are totally positive that you have never known anyone called Brian?"
> *"Well I'm not completely sure, I can't think of anyone by that name though."*
> "But you're not positive? Well that's definitely the name I'm getting through so have a think about it and see if anything comes to you."

Many of the techniques used in the win win game work in a similar way to this one. The implication is always that the psychic is right and that it only seems that they are wrong due to some flaw on the subjects part. Maybe the psychic is right and the subject has simply forgotten, or perhaps the psychic is right and the subject just doesn't know it yet, in fact maybe no one knows yet. Another alternative is to claim that the information is embarrassing in nature and that is why the subject doesn't know about it as their departed loved one chose to keep it to themselves.

Of course the alternative to this is for the psychic to admit that they are wrong, but that it isn't really a problem. Again they can do this a number of ways. The psychic might admit that they were wrong but that it doesn't matter because it was just an incidental detail anyway. They might claim that they are wrong now but that in the future they will be shown to be right or that while they are wrong in a factual sense that they are correct in an emotional sense. And of course they can always just admit that they got it wrong, remind the subject that they never claimed they would get it right every time and just move on. Whatever the method the psychic still wins.

But at the end of the day it doesn't really matter how many hits or misses the psychic scored during the reading itself as long as they can leave the subject with the sense that the reading was a success. And by far the best way to do this is simply recap the points that went well and ignore those that failed to bear fruit. Looking back at some of the examples I used so far we can see how this would work. For example when we looked at fuzzy facts our psychic made a very vague and general statement.

> "I'm sensing some connection to America or the Americas."

However when it comes time to recap the psychic can plant the idea that they were specific from the beginning.

> "I sensed that you are planning to go to Canada to see the Calgary Stampede."

Of course this is far from what was actually said, but a good psychic will do her best to make sure that the subject remembers it that way. Another example might be the one I used in the vanishing negatives section. While in reality the psychic asked the subject if they worked in a bank and then implied that they had known the answer all along, in the recap they can simply state outright that they knew what the subject did for a living. Most psychic readings are far more impressive in hindsight than they are at the time.

But of course I have so far ignored a large part of any psychic's repertoire, something I myself included in my reading, and that's the psychic prediction. Now by this point you won't be surprised to learn that there are many different types of prediction that a psychic can choose from. Let's start with the type that I used in my reading.

The Unlikely Prediction

The type of prediction I used in my reading is the one with the lowest chance of coming true but the biggest pay off if it does. To remind you, this was my prediction.

> "A week after finishing this book you will bump into someone who you haven't seen in a while who will tell you about a big change that is about to happen in their life that relates to their job."

That's a pretty specific prediction. Not only do I say what will happen but I also set a time period in which the event will take place. Now while there is still plenty of room for interpretation, after all I don't say specifically who you will meet or what exactly the big change will be, only that it relates to their job, this is still a prediction that will either happen or not and as such has a very high chance of being wrong. So why would any psychic make a prediction like this?

Well it all comes back to what we learnt in the chapter on the law of large numbers. Even though this prediction is unlikely to come true for most of you if I give the same prediction to enough people eventually something close enough to the event I describe with happen within the time scale allotted and you can bet for that person this will appear to be an amazing confirmation of my psychic powers. That person then runs off and tells their friends about how amazingly accurate my predictions are and new business comes flooding in. As for everyone else, well most people don't spend much time trying to remember things that didn't happen. Most people will have forgotten the prediction I made by the time they finish reading this book. Only if the prediction comes true will it pop into their minds, otherwise it will most likely be gone forever.

The Likely Prediction

Of course the opposite of an unlikely prediction is the likely prediction and, as the name implies, this is a prediction that is almost guaranteed to come true in some way or another. Let's take the example from my own reading and turn it from an unlikely into a likely prediction.

> "Within the next week or so you will bump into someone you haven't seen in a while who will tell you news about something happening in their life."

That is basically the same thing I said in the unlikely prediction, only now there is much more scope for it to come true. The time period is much more vague for a start. How many weeks does "the next week or so" cover? Does that include this week as well? Also I've been much more general about what the person will say. Rather than them specifically telling you about big news related to their job now I'm just predicting that they will tell you news about something in their life which, unless they have gone mute since you saw them last, is pretty much a given. But don't let me give

Building your Skeptical Toolkit

you the wrong impression that likely predictions can't sound specific, because they can. For example:

> "In the year ahead I can see that you will be involved in an accident involving water that will leave you upset but ultimately unharmed."

At first glance this sounds fairly specific, but is it really? What exactly do I mean by "in the year ahead"? Do I mean before the end of the calendar year or within the next 365 days from now? The term accident is equally nebulous and could cover everything from a minor mishap to a fatality, and what exactly do I mean when I say that you will be "involved" in it? Does that mean it will happen directly to you or that you will simply witness it or know the person it happens to? "Upset but ultimately unharmed" covers a wide range of possibilities also. Do I mean that you will be unharmed by the accident or that, after a period of recovery, you will eventually be alright? And finally saying it will involve water covers everything from being splashed by a passing car in your favourite outfit to breaking your umbrella in the rain to almost drowning. There is so much scope for interpretation in that short sentence that I am practically guaranteed a hit of some sort.

One-way Verifiable Predictions

There are many other types of prediction that a psychic can use but this is the last one I want to look at here. A one-way verifiable prediction is a prediction that can only be verified if it comes true. If it doesn't come true then there is no way for the subject to ever prove that it didn't, thus making it a very useful weapon in the psychic's arsenal. An example of this might go like this:

> "A friend may decide to call with an offer of a potentially lucrative investment opportunity, though they may decide at the last minutes that it sounds too much of a gamble and not do so."

The basic formula of the one-way verifiable prediction is always the same. Some other person is planning on doing something that relates to you, however they may decide not to do it after all. If someone does contact you with an investment opportunity then of course the psychic can claim that her prediction came true. However if they don't contact you then the psychic can still count that as a hit, after all she did say that they might not do so. In fact the only way to prove this sort of prediction wrong is to go round to all your friends and ask them if they'd been thinking of contacting you with an investment opportunity, but even then the psychic could simply claim that the person in question had lied to you and that is why they said no. Simply put, this sort of prediction cannot fail for the psychic.

Anyway that is all I'm going to say on the subject of cold reading, even though I have only covered a fraction of the techniques available to a skilled psychic. If nothing else I hope I have conveyed to you that there is a lot more to cold reading than just making vague guesses and telling people what they want to hear. A skilled cold reader really can appear to know things about you that they really shouldn't know by natural means and even if you know what is going on the effect can be impressive and somewhat unnerving.

There is one last thing I want to touch on in this chapter and that is the flip side of the cold reading coin, hot reading. Hot reading is where the psychic knows some information about you in advance and simply presents it to you as though they are coming about it by psychic means. Hot reading techniques are often mixed in with the various cold reading skills used by the psychic. For example when the subject books an appointment the psychic might casually enquire who they hope to make contact with. Then on the day of the reading itself the psychic will simply feed this information back to the subject dressed up in the paraphernalia of the psychic reading. It is also not uncommon for psychics who do public readings rather than one on one sessions to have regulars who come along to many of their shows. These are people the psychic has had a relationship with often for years and about whom they know a great deal. If the show is not going too well the psychic can always call upon one of these familiar faces and rattle off a number of amazingly accurate details about that person's life much to the delight of the crowd. Of course what they are doing here is hot reading, simply feeding back information that they have gained at an earlier date by other means under the pretense of psychic powers.

And these days with the rise of social networking the psychic's job is just getting easier. If the psychic knows in advance who they will be giving a reading to, which is often the case, then a quick Google search or visit to their Facebook page can reveal a wealth of information, especially if the person is any sort of celebrity. In another of his 2008 programmes on the paranormal Tony Robinson investigated the subject of psychic readings and after getting a reading himself stated that nothing the psychic had said about him couldn't have been discovered in five minutes on the internet.

Now I am not saying that psychic powers don't exist and that all psychics use the various techniques that I have outlined over the previous pages. It is of course entirely possible that some psychics genuinely have powers that allow them to contact the dead or divine details about the future. However, to misquote magician

Your Brain Hates You!

James Randi, while they may have psychic powers by means of which they can read our minds; if so, they appear to be doing it the hard way.[*]

---

[*] James Randi's original quote relates to "psychic superstar" Uri Geller and correctly goes like this: "Uri Geller may have psychic powers by means of which he can bend spoons; if so, he appears to be doing it the hard way."

Skeptical Sidebar #7
The book test

So we all know that you can't actually read someone's mind, but there are plenty of ways to make it look like you can. Mentalists and magicians do this all the time in their shows and one of the methods they like to use is known as the book test. Now there are probably as many versions of the book test as there are magicians performing it but they all have pretty much the same basic structure. You are asked to select a book and then the magician gets you to pick a page and a word on that page apparently at random. You are asked to concentrate on that word and then, after a moment for dramatic effect, the magician reveals the word you were thinking about, much to the amazement of everybody. Ok so pay attention as I am going to teach you one of the many ways in which this can be done.

For this trick, sorry I mean mind reading extravaganza, you will need three average sized books, some paper and a pen. We are also going to make use of the cold reading trick of getting people to remember what you want them to remember, rather than what actually happened. So here is what you need to do.

Your Brain Hates You!

First start by introducing your subject to what you are going to do, or at least what you want them to believe you are going to do. Tell them that you are going to get them to pick a word, completely at random, from one of the three books and that you are then going to read their mind and tell them what that word is. Show them the books, make it clear that they are real books, that all the pages are different and that there are lots of possible words for them to choose from. Next have them select one of the books at random. Make it clear that you are not influencing their choice and that they are free to change their mind if they feel you have in anyway. Once they are happy that they have freely selected a book, place the other two to one side and hand them the pen and paper.

Explain that you want to make sure that they pick a completely random word and do so in as fair a way as possible, one that you can have no influence over. In order to do this ask them to write down any three digit number across the top of the paper and to make sure that all the digits are different. Look away and make sure they hide it from you as well so that you can't know what number they are writing. For this example let's say they pick...

742

Now get them to reverse this number and write it underneath the first number they picked. If you have more than one person watching the trick you might want to get someone else to do this, you know to rule out the possibility that the first person is in on the trick. Again make sure that you can't see what they are doing.

742
247

312

Building your Skeptical Toolkit

Now have them subtract which ever number is lower from which ever number is higher and write down the answer. This will give them a new number.

742 –
247
495

Once again get them to flip this number around. If the number is less than three digits get them to put zeros at the start before flipping it to make it up to three digits, so that for example 12 becomes 012 which becomes 210. Once they have done this get them to add these two numbers together to get the final, completely random number.

742 –
247
495 +
594
1089

So now we have the number we are going to use to help them select a completely random word from the book that they freely choose from the three on offer. Tell them to keep this number hidden even now and to circle the last digit in this number.

108(9)

This circled digit becomes the word number and the other three digits represent the page number. Now, making sure you can't possibly see, get them to open the book to that page and counting from the first word at the top of the page find whatever word is located at the position indicated by the circled number. Get them to remember this word and then close the book. Remind them that they freely selected which book to use and that they then came up with a completely random number that you in no way influenced them to select and that they then flipped, subtracted, flipped again and then added to get a new number, the value of which not even they could have guessed at. Finally they used this completely random number to find a single word from the thousands in the book before them. Now have them really concentrate on the word, get them to say it over and over again in their mind. Focus on their eyes and then, after a few seconds, take the pen and a fresh bit of paper and write down the word they are thinking of. Fold the paper tightly in half and place in the middle of the table for all to see. Ask them to tell everyone the word they were thinking of, let out a deep breath and have them unfold the piece of paper and read out the word you read from their mind. Bask in the applause.

But wait, I bet I can read your mind right now. You're thinking "Ok, but how they hell do I know what word to write down?" Am I right? Well here's the trick. Let's imagine we are doing this trick for three different people at different times. We get each one of them to select a book at random and to come up with a three digit number completely at random. We then have them do all the flipping and adding etc to generate a new number. Here's an example of what we might get. Remember you subtract the lower number from the higher one each time.

271 −	329 −	821 −
172	923	128
099 +	594 +	693 +
990	495	396
1089	1089	1089

As you can see it doesn't matter what three digit number you start with as you always end up with the number 1089. Now it is just a case of you memorizing the 9th word on the 108th page of all three books. Keep reminding them that they picked the book and number completely at random, which is true, and selected a word that you couldn't have known in advance, which is not true, and they will remember this being the case and tell all their friends of your amazing mind reading powers.

Your Brain Hates You!

Building your Skeptical Toolkit

Can I have a vowel please Carol?
Devils, Dowsing and the Ideomotor Effect

It's late on a cold winter's night. Outside the wind is blowing and rain is falling in sheets. Inside you and a few of your close friends are huddled around the fire trying to keep warm. The storm has knocked out the power so candles light the room and with no TV you have to find some other way to entertain yourselves. You remember that, packed away with your Monopoly and Trivial Pursuit sets, you have an old Ouija board that someone gave you as a joke gift years ago and decide that now might be the perfect time to take it for a spin. You and your friends gather around the board, each placing one finger on the planchette, and with a nervous giggle you ask if anyone is there. At first nothing happens, but then the planchette slowly starts to move across the surface of the board and comes to rest on the word YES. You all look at each other in surprise. One by one you deny being the one to have moved it. It seems that the planchette may have really move by itself. With a chill running down your back you ask the board who is it that you are talking to. Slowly, but gaining speed with each letter, the planchette spells out the reply. D...E...V...I...L. At which point you all freak out.

Whether you've had any personal experience with a Ouija board or not the above story will most likely seem familiar to you. A group of people gather around an ornate board upon which the words 'Yes', 'No', 'Goodbye' and 'Maybe' are laid out around the twenty-six letters of the alphabet and the numbers zero to nine. A heart shaped planchette, often made of wood or glass, though more often these days of plastic, sits on the board and each person places a single finger gently upon it as one member of the group offers up questions into the silence. Suddenly the planchette moves and everyone lets out a cry as they all know that they were not the one to have moved it.

However, despite what the movies and late night TV evangelists may have told you, there is nothing spiritual or supernatural about the Ouija board, and the forces behind the movement of the planchette have a much more mundane and yet interesting cause than ghosts and devils from the other side.

The Ideomotor Effect, first identified by psychologist William B. Carpenter in 1852, is a psychological phenomenon in which a person makes involuntary or unconscious movements, most commonly of the hands and arms, in response to some suggestion or the expectation of something happening. Now if you haven't come across this idea before then the thought that your body could be moving and doing things without your conscious input may sound a little hard to accept. However, if you think about it there are many occasions in life where our bodies do things without us explicitly telling them to. Perhaps you find that you get tears in your eyes at particularly emotional moments in movies or maybe you are one of those people for whom yawning is particularly contagious. And then of course there are those times when you accidently touch something hot and your body reacts upon reflex to stop you from getting burnt, often before you are fully aware of what is going on.

The ideomotor effect has a lot in common with these more everyday involuntary reactions, known as sensorimotor actions, and is an important phenomenon for skeptics to know about as it provides by far the most plausible explanation for all manner of pseudoscientific and paranormal claims. Automatic writing, dowsing, facilitated communication, mathematical horses and even alternative medical modalities like applied kinesiology can

Building your Skeptical Toolkit

be, fairly conclusively, shown to be the result of the ideomotor effect in action. We'll take a look at each of these areas in turn but seeing as I have already mentioned the Ouija board let's clear that one up first.

As I said the ideomotor effect is the body reacting unconsciously to stimulation such as suggestion and expectation, and nowhere is this clearer than with the Ouija board. No matter whether someone is trying it for the first time or is a long time user, there is an expectation that something will happen when you use a Ouija board and it is this expectation that allows the ideomotor effect to take place. The people playing the game honestly believe that they are not moving the planchette around and, once you have eliminated everyone at the table, it is not uncommon for them to attribute the movement to "spiritual forces". People ask questions and it is the expectation of an answer that causes them to get one rather than any real communication from the other side.

When it comes to spelling out words expectations again play a big part. In the story at the beginning of the chapter the people playing the game found themselves apparently getting a message from the Devil, but let's think about that for a second. People tend to think that Ouija boards are in some way evil, thanks in no small part to various religious and spiritualist organisations saying so numerous times over the years. Now the first letter they came up with could have been completely random, but once you have that D your mind can easily start to fill in the rest. With every letter after that you become more and more convinced that you know what it is going to spell and so enforce the power of the ideomotor effect. By the time you get to the V everyone's unconscious is singing from the same song sheet and the rest of the message is spelt out with increasing speed.

319

Your Brain Hates You!

On top of this it is possible that those playing the game may find their deepest thoughts, fears and fantasies spelt out before them, producing messages that it is far more comfortable to attribute to evil spirits than to one's own subconscious. If you are expecting a disturbing message then your mind will be only too happy to fill in the gaps. This spelling out of messages that differ from the thoughts in the players conscious mind can be a powerful incentive to believe that something mystical is going on. However when getting a message about having sex whilst eating cake and killing puppies* you should be much more concerned about the person across the table from you than any ghost or demon.

Going off on a slight tangent for a moment but it is rather telling, and not just in relation to Ouija boards here, that when psychics or mediums claim to make contact with the spirits of ancient people they always talk and spell in perfect 21st century English. You never get a group of people who don't speak, for example, German and who know nothing about the history of language using the planchette to spell out perfect 16th century German sentences. Similarly jumping back to the subject of past life regression you never come across someone who claims to have been Julius Caesar in a previous life suddenly being able to speak in perfect Latin. It just doesn't happen because in both cases the answers come completely from the minds of those involved and not from any real connection with the spirit world.

However it is possible you remain unconvinced. Maybe you have used a Ouija board yourself and find the idea that you were the one spelling out the answers difficult to swallow. In that case there is a very simple experiment, used to great effect by skeptical magicians Penn and Teller on their TV show Bullshit!, that you can try for yourself. On their show Penn and Teller had

---

* My thanks to Youtuber and fellow League of Reason blogger Th1sWasATriumph for this delightful bit of vileness.

Building your Skeptical Toolkit

three strong believers in the paranormal powers of the Ouija board get together and ask a series of yes or no questions to the spirit world, the answers to which were given with uncanny accuracy via the planchette.

Penn and Teller then blindfolded the three participants and, without their knowledge, turned the board around 180 degrees. As before they asked questions of the spirits and this time the planchette travelled to bare areas of the board where the participants believed the "Yes" and "No" marks were located. Ever the skeptics Penn and Teller ended their experiment with the question of why the spirits would have been affected by this simple trick and the blindfolding of the people holding the planchette if it was truly them moving it around. I will let you make up your own minds on that one.

I'm going to leave the topic of automatic writing until last as it segues nicely into the topic of the next chapter. Instead we're going to move on and take a look and the subject of dowsing. Dowsing, in its most basic and recognised format, is a form of divination that claims to be able to find water using a Y or L shaped twig or rod held in the hands of the dowser. These rods move, apparently of their own volition, and guide the dowser to the location of the water. These days modern dowsers use a wide variety of different versions of this idea and claim to be able to find, well, pretty much anything with them.

The apparent effect of dowsing, despite its continued widespread popularity, has been repeatedly shown in test after test to be the result of purely natural explanations such as confirmation bias and, of course, the ideomotor effect. The overwhelming amount of evidence against the validity of this particular paranormal claim can perhaps be explained by the ease by which the claims can be tested. A dowser who claims to be able to find water with an 80% level of accuracy can be tested simply with a few upturned

321

buckets, beneath a number of which is placed a glass of water. If the dowser really does have the power they claim then it would be expected that they be able to identify which buckets have glasses of water beneath them with a success rate far above that predicted by chance alone.

A 1948 study published in the New Zealand Journal of Science and Technology tested 58 dowsers and concluded that none of them performed any better than chance. Magician James Randi conducted a similar experiment with a number of proclaimed dowsers as part of his then $10,000 challenge, which has now evolved into the much more impressive sounding Million Dollar Challenge. A series of underground pipes were laid out and it was agreed that if any of the dowsers taking part could locate which ones had water in them with an 80% success rate then they would get the money. All the participants failed, despite having claimed years of success at locating water under similar conditions.

It would be remiss of me not to mention the 1987-1988 "Scheunen" experiments, conducted by Hans-Dieter Betz and a number of his colleagues in Munich, which provide the strongest evidence in support of dowsing. Starting initially with 500 dowsers numerous experiments were run to identify the 43 most skilled amongst them. Over the course of two years hundreds of additional tests were run, during which it was concluded that 37 of the preselected 43 had no discernible dowsing ability. The remaining 6 participants however appeared to demonstrate *"an extraordinarily high rate of success, which can scarcely if at all be explained as due to chance"*.

However in a later review of the data, published in Skeptical Inquirer, professor of physiology and a leading skeptic Professor Jim Enright concluded that the experiments provided *"the most convincing disproof imaginable that dowsers can do what they claim"*. In part his argument addressed the fact that the results

Building your Skeptical Toolkit

were not reproducible, firstly inter-individually in that of the 500 proclaimed dowsers only 6 were believed to demonstrate any level of ability above that of chance. And secondly intra-individually, in that of the remaining 6 their abilities appeared to vary from test to test with them succeeding well at a test one day and the next performing no better than the rest of the 500 "unskilled" dowsers. It seems likely that some form of confirmation bias was at work during these experiments and Enright himself had this to say on the matter:

> "The researchers themselves concluded that the outcome unquestionably demonstrated successful dowsing abilities, but a thoughtful re-examination of the data indicates that such an interpretation can only be regarded as the result of wishful thinking."

To my mind one of the clearest indicators that it is the ideomotor effect at work rather than any real paranormal power can be found from simply observing the actions of both novice and "professional" dowsers. A novice dowser picking up the rods for the first time will find them crossing when they pass over an area where they have been told water can be found, even, and this is the important part, if no water is actually located at that spot. In 1992 psychology professor Ray Hyman, acting as a consultant and expert witness for the State of Oregon, conducted an experiment using two groups of student volunteers. He started by demonstrating how dowsing rods worked by walking slowly across the room until the rods suddenly crossed. He then explained that the reason they had crossed was because there was a water pipe under the floor at that point. The students then had a go and sure enough the rods crossed when they passed over the point that Hyman had indicated.

He then repeated the experiment with the second group and once again got the same results. However this time around he picked another random point in the room at which to cross the rods and once again gave the same explanation as to why they had crossed there. As before the students found the rods crossing at the point Hyman had indicated to them and nowhere else. As well as this experimental evidence for the ideomotor effect many videos can be found of dowsers in action in which it is clear that the movement of their hands and arms is what is actually causing the rods to cross and not some magical power.

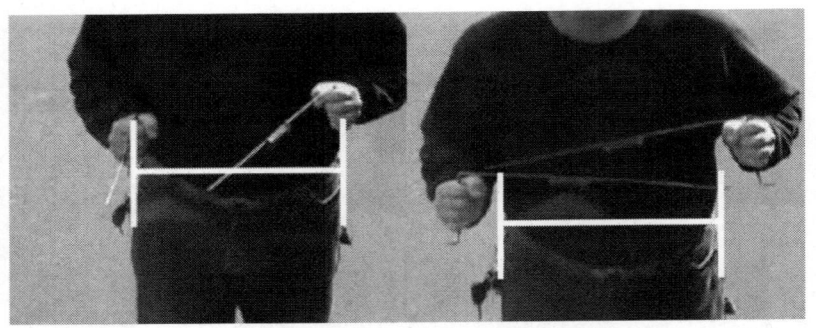

Two frames from a video application to the James Randi Million Dollar Challenge taken less than a second apart and accompanied by the words "You can clearly see I do not do that." Er, not so much.[*] The white bars indicate the fixed width of his hips as a comparison to the non-fixed position of his hands.

Of course it is important to remember that the person holding the rods may be completely unaware that they are responsible for them moving rather than some outside force. It can honestly feel as though the rods moving in their hands are doing so of their own accord and so we shouldn't jump to the conclusion that they

---

[*] For the complete video see the James Randi Foundation channel on youtube and decide for yourself if mystical powers or the ideomotor effect is at work. www.youtube.com/user/JamesRandiFoundation

Building your Skeptical Toolkit

must know what they are doing. Nor should we assume that they are somehow less intelligent or foolish for failing to notice that they are the one responsible for the movement. After all that is pretty much the entire point of the ideomotor effect, that you don't realise that you are doing anything.

Before we move on I want to quickly take a look at the dark side of dowsing, something I touched upon in the introduction to this book. The ADE651, produced by the UK based company ATSC, is a hand held device that, according to the promotional material, uses *"electrostatic magnetic ion attraction"* to enable the operator to detect the presence and location of various drugs, explosives and other substances at a distance of up to a kilometer at ground level and five kilometers in the air. By programming the detection cards *"to specifically target a particular substance, (through the proprietary process of electro-static matching of the ionic charge and structure of the substance), the ADE651 will "by-pass" all known attempts to conceal the target substance. It has been shown to penetrate lead, other metals, concrete, and other matter (including hiding in the body) used in attempts to block the attraction."* Wow, that is pretty impressive and if accurate then it more than justifies the price tag of between £11,500 and £37,000 that ATSC charged per unit. And it seems that the Iraqi government certainly thought so, spending £52m over two years for 1,500 units. The ADE651 units were deployed by the Iraqi Police Service and the Iraqi Army at hundreds of police and military checkpoints across Iraq, often being utilised as a replacement for more conventional inspection methods. There was only one small problem, these things simply don't work.

In late 2009 a guard and a driver working for The New York Times, both of whom were licensed to carry firearms, passed through nine police checkpoints that were using the ADE651 device. Not a single one of the checkpoint guards using the devices detected the two AK-47 assault rifles and ammunition that the two men had inside their vehicle. But of course this could have been a one off event, maybe there was a specific reason why these particular weapons passed through undetected? Well not according to the United States Military and numerous technical experts who have examined the devices and concluded that they are completely useless.

But what has an apparently high tech device like this, whether it works or not, got to do with the ideomotor effect. Maybe if I explain how they are used it will all become clear. The ADE651 consists of a plastic handgrip upon which is mounted a swiveling telescopic antenna. Various plastic coated cardboard detection cards, each one specially prepared to detect specific things, can be slotted into the device and you are all set to go. The ADE651 requires no batteries or power source as it draws its energy directly from the body of the operator. The operator holds the device at a right angle to their body and slowly walks the area. If there is a bomb in the area that antenna will swivel to point in the direction of the explosive. Sound familiar?

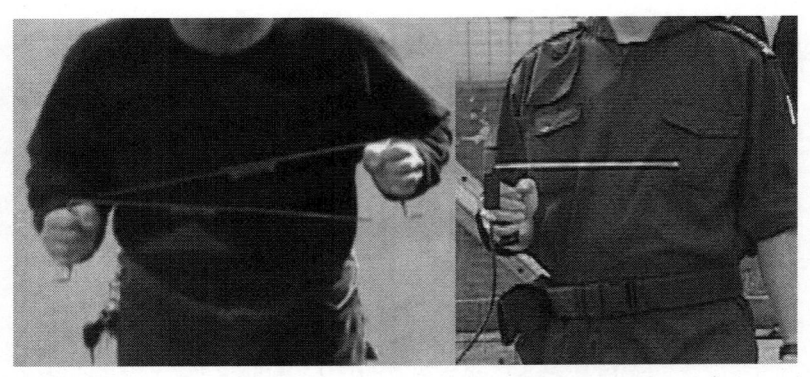

Traditional dowsing rods on the left and the ADE651 on the right. Am I the only one struggling to see the difference?

Building your Skeptical Toolkit

As retired United States Air Force officer Lt. Col. Hal Bidlack put it, the ADE651 is nothing more than an explosives divining rod. Maj. Gen. Richard J. Rowe Jr., who was at the time responsible for overseeing the Iraqi police training for the American military, added that he doesn't *"believe there's a magic wand that can detect explosives, if there was, we would all be using it. I have no confidence that these work."* Furthermore reporters for the BBC demonstrated that the "programmable detection cards" contained nothing more advanced than security tags used to stop shoplifting and certainly nothing that could be classed as programmable.

The Iraqi government it seems fell victim to the ideomotor effect, as well as what could be reasonably classed as less than honest business practices. The UK has now banned the export of these devices and the managing director of the company, Jim McCormick, has been arrested on suspicion of fraud. But as I said in the introduction these action appears to have come a couple of hundred potentially preventable deaths too late.

Well that was a bit of a downer, what say we lighten things up a bit? And what's more effective at lightening things up than Clever Hans, the mathematical horse? For those of you who have not heard the story let me run through it quickly for you. The year is 1891 and German William von Osten has begun showing off the amazing abilities of his Orlov Trotter horse, known as Clever Hans. Von Osten claimed that Hans was able to not only understand human language but to grasp mathematical concepts as well. Von Osten would ask the horse a question, such as what is three plus two, and Hans would correctly respond by tapping his hoof five times.

Because of the large amount of interest from the public in this matter, the German board of education established the Hans Commission, a team of 13 people including a veterinarian, a circus

327

manager, a Cavalry officer, a number of school teachers, and the director of the Berlin zoological gardens, to investigate Von Osten's claims. In 1904, after "careful" examination the commission concluded that there were no tricks involved in Clever Hans' performance. So was Clever Hans really a horse that could do maths?

Clever Hans in action.

Unfortunately, as much as I wish it was not, the answer is no. Enter psychologist Oskar Pfungst, who took over the investigation from the commission and promptly set up a number of tests to see if Hans could really do what he appeared to do. These tests were relatively simple and included separating Hans from both Von Osten and the audience so that he could only hear the questions, having someone other than Von Osten asking the questions and asking questions that Von Osten himself did not know the answer to.

Pfungst quickly discovered that, while it didn't matter who asked the question, Hans only appeared to get the correct answer when he could both see the person asking the question and the questioner themselves knew the answer to the question. So

Pfungst turned his attention to the questioners themselves and quickly found that they were unconsciously indicating to Hans when he had reached the right answers, at which point Hans would stop tapping. As Professor Ray Hyman puts it *"Hans was responding to a simple, involuntary postural adjustment by the questioner, which was his cue to start tapping, and an unconscious, almost imperceptible head movement, which was his cue to stop."*

The story of Clever Hans is yet another example of the ideomotor effect at work, and yet one with far reaching implications in the scientific world. Pfungst's work showed how animals can pick up on subtle clues from humans and this knowledge has helped shape investigation into animal cognition ever since. Furthermore Pfungst's findings, which became known as the Clever Hans effect, also went on to provide important insight into another psychological effect known as the observer-expectancy effect. The observer-expectancy effect, as the name implies, is where the subject of a test picks up clues as to what results are hoped for from the examiner themselves. This is why, as we looked at earlier in the book, double-blinded testing is used in medical trials in order to make sure that the person being tested cannot pick up any clues as to what results are expected by the person administering the treatment under examination.

So as we are talking about medicine let's take a look at Applied Kinesiology, a diagnostic method used in alternative medicine. So what exactly is applied kinesiology? Well firstly it is important to note that it is not the same thing as kinesiology, which is the scientific study of human movement. Applied kinesiology on the other hand, according to the International College of Applied Kinesiology, is *"a system that evaluates structural, chemical and mental aspects of health using manual muscle testing with other standard methods of diagnosis."*

Your Brain Hates You!

Well that doesn't sound too bad, in fact it sounds pretty scientific. However it is worth noting that the central premise upon which applied kinesiology is based, that being the idea that problems with internal organs, such as the heart or liver, are always accompanied by a weakness in a specific corresponding muscle, is not supported by the evidence or upheld by current mainstream medical theories. Additionally the underlying reason for the belief in this unsupported connection is itself based upon the equally unsupported concepts of chiropractic and certain aspects of traditional Chinese medicine, namely that blockages in the body's energy flow, often referred to as Qi or innate intelligence, are responsible for most if not all forms of sickness. Again these claims are not supported by the findings of medical science.

So with that in mind how is applied kinesiology said to work? Well by way of a few manual muscle tests the practitioner can measure the resistance and smoothness of certain muscle groups and by identifying weakness in specific areas can diagnose issues with the organs to which those weak muscles are said to be linked. They also often test for problems with balance and flexibility as well as the effect that assorted chemicals, most often placed upon the tongue or simply held in the hand, have upon these responses.

This last aspects generally works in two ways, with positive nutritional or chemical supplements being used to strengthen previously weak muscles and exposure to allergens being used to weaken previously strong muscles in order to test the patient's response to potentially harmful or imbalancing substances that may be the cause of their sickness. The practitioner is then able provide specific products or treatments in order to strengthen your weak muscles and/or protect you from the various allergens to which you are most susceptible. One of the most common examples of the kind of treatments on offer, and the one that you are most likely to be familiar with, are copper bracelets that are

Building your Skeptical Toolkit

said to help restore balance and improve the flow of energy within your body and which are often used as a treatment for arthritic pain.

So what has all that got to do with the ideomotor effect? Well I'm glad you asked, and let me answer that question by way of an actual test used by practitioners of applied kinesiology that is said to measure everything we have discussed so far, from muscle resistance and balance to the patient's response to positive chemical supplements. Go grab yourself a friend or family member so that you can play along at home.

The patient is requested to hold their arms out straight on either side of their body and keep them rigid. They are then instructed to stand on one foot and do their best to keep their balance as the practitioner applies downward pressure to the mid-point of the arm on the side of their body on which they have raised their foot. They do their best but even with very little pressure applied they quickly find themselves losing balance.

Next the practitioner asks them to hold an item, or places a drop of some nutritional supplement on their tongue, that is said to counter the harmful and imbalancing effects of allergens in the environment. The test is repeated and this time, despite the practitioner bringing considerable more effort to bear, the patient is able to keep their balance. The patient is adamant that they tried equally hard to keep their balance both times and as the only apparent difference between the two tests was the item or supplement given to them by the practitioner they find themselves convinced by the claims of applied kinesiology.

By now you know enough about the ideomotor effect to know that expectation can play a massive role in things like this. Despite what the patient may believe about their efforts to retain their balance, the expectation that the magical doohickey will have an effect is often enough to make it so. However in this case it is not

331

only the patient that is falling victim to the ideomotor effect as the practitioner, assuming of course that they are honestly unaware of their actions, can be equally susceptible. Here's how you can recreate the compelling effects of this test using the most mundane of household items.

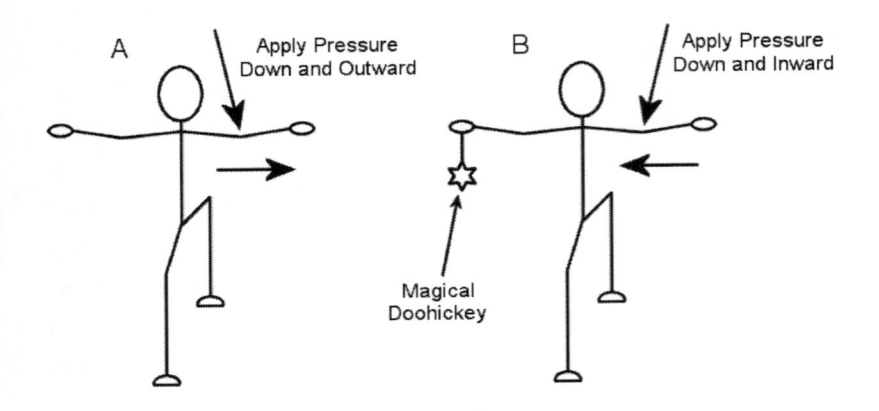

```
         A                                    B
       Apply Pressure                      Apply Pressure
       Down and Outward                    Down and Inward

                    (stick figure A)              (stick figure B)
                                    Magical
                                    Doohickey
```

Carry out the test as described above, getting your friend to hold out their arms and stand on one leg. Apply a gentle but constant pressure to their arm in a downward and slightly away from their body direction (see picture A above). Very quickly and with little effort on your part you will find them losing balance. Now get them to hold something in the hand on the opposite side of their body to where you applied pressure. It really doesn't matter what it is, anything lying around your house will do. I have done this test with a bent fifty pence piece and a packet of sweetener and have got the exact same results. However, if you can find something a little unusual looking then your victim, umm I mean participant, will be much more inclined to believe that the item was responsible for the change[*]. With them holding the "magical"

---

[*] A great thing to use would be a Placebo band, created by the SkepticBros to look like actual applied kinesiology devices, but without all the accompanying pseudoscience - http://skepticbros.com/

item get them to once more stand on one leg and this time apply pressure to their arm in a downward and towards their body direction (see picture B). You will find they you are able to apply a great deal more pressure than before and yet they will still be able to keep their balance.

It is completely plausible that, due to the ideomotor effect, genuine believers in applied kinesiology are totally unaware that they are applying the pressure in a different manner each time round, and of course from the point of view of the patient it all feels very convincing. The vast majority of the tests involved in applied kinesiology work in a similar manner to the one described above, with pressure applied in a slightly different manner each time producing dramatically different results. Reversing the test, so that the patient starts by holding some alleged allergen only for it to be removed for the second phase of the test, is equally compelling.

Ok so we are almost finished with the ideomotor effect, which by now I am sure you can see is a simple but powerful thing indeed, and I am sorry to say that I am going to have to bring things down again as we look at the subject of facilitated communication. To my mind facilitated communication is probably the most harmful way in which the ideomotor effect comes into play, as at its best, and when performed by well-meaning people who genuinely want to help, all it can offer is false hope, while at its worst it can ruin the lives of entire families.

Facilitated communication is the idea that impaired individuals who are unable to communicate clearly by way of speech or sign language, such as those with severe autism, mental retardation or those suffering from brain damage, can type out messages on a keyboard, or similar devices, with the help of a facilitator supporting their hand or arm. The facilitator, their hand placed over that of the patient or upon their wrist or arm, guides the patient's hand over the letters of a keyboard or the words and symbols on a board or touch screen. The patient then allegedly communicates to the facilitator which letter or symbols they wish to press, thus spelling out messages containing their thoughts and feelings, as well as the occasional bit of poetry.

The appeal of facilitated communication is not at all hard to understand. Being the parent of a child that is unable to communicate with you must be incredibly hard. Discovering a method by which your child is, perhaps for the first time, able to tell you that they love you and that they are happy must be emotionally overwhelming. Furthermore, finding out that, beneath their inability to communicate, your child is intelligent, maybe even beyond their years, is surely something that every parent of an autistic or otherwise mentally disabled child dreams of. It is hard to fault someone in a situation like that for allowing their hope, and their need for this assisted form of communication to be real, to override their critical thinking skills.

I also find my feelings incredibly conflicted when it comes to the facilitators themselves. On the one hand the evidence from numerous investigations into the subject, including those carried out by the American Psychological Association as well as psychologists at Harvard University, points to the fact that facilitated communication is just another example of the ideomotor effect at work and that the messages undoubtedly come unconsciously from the facilitators themselves, rather than the patient, based upon their own expectations as to what their patient wishes to relay. On the other hand there is the fact that the vast majority of those involved in facilitated communication genuinely believe that they are communicating with these otherwise uncommunicative people and want to use this ability for the betterment of both the patient's life and that of their family. These people are not charlatans, they are good, caring

Building your Skeptical Toolkit

people who just want to help and believe they have found a way in which they can.

And if that was all there was to facilitated communication then I would probably file it away under the heading of "ultimately pretty harmless nonsense" along with mathematical horses and Ouija boards. Unfortunately however this is not the case as, while most facilitators do simply pass on messages of love and happiness, there are those times when things take a dark turn and the facilitators relay tales of torment and abuse. During the 1980s and on into the early 1990s a number of people were convicted in cases of alleged child abuse based upon testimony given by way of facilitated communication. Like the abuse cases we looked at when discussing the subject of repressed memories these unconventional testimonies were generally accepted uncritically and treated as firsthand accounts from the alleged abuse victims themselves. People have gone to jail and families have been torn apart based upon a methodology that the American Psychological Association describes as *"a controversial and unproved communicative procedure with no scientifically demonstrated support for its efficacy"* noting that *"[studies] have repeatedly demonstrated that facilitated communication is not a scientifically valid technique for individuals with autism or mental retardation."*

In fact as recently as 2008 facilitated communication was accepted as a legitimate form of testimony in a child abuse case in Oakland Michigan. The parents of a 14 year old nonverbal autistic girl fell victim to the dark side of facilitated communication, having already been convinced of its legitimacy after apparently receiving insight into their daughter's inner thoughts and feelings by way of a facilitator. When Cindi Scarsella, a teacher's aide at their daughter's school who it seems had received only hours of training in the technique, conducted a facilitated communication session on the girl she immediately "discovered" that the child had, without the mother's knowledge, been raped by her father on and off over the previous eight years. The school contacted the police and the father was promptly arrested, charged and imprisoned while the case awaited its time in court. The child was also removed from the rest of her family and place in care pending the outcome of the case.

Ultimately the charges were dropped, the father released and the girl returned to her family, but only after facilitated communication itself was put on trial and shown, once again, to be an invalid form of communication. The attorneys and experts acting for the defence proposed a series of tests in which questions were asked with the facilitator out of the room. The facilitator was then brought in to record the answers to the questions asked in their absence. The following outcome was reported:

> The prosecution's case suffered additional difficulties in the January hearing when FC failed to produce a single correct answer in two separate tests. In these tests, the facilitators had not been allowed to hear the simple questions asked of the girl. The facilitators had testified that it was not necessary to hear the questions to successfully "facilitate" successfully. Defense experts had testified that scientific evidence showed that FC involves complete control of the output by the facilitators, and had predicted that FC would fail when the facilitators could not hear the questions.

But let's leave the depressing subject of facilitated communication and end the chapter with something a little more upbeat. Automatic writing is a form of writing that proponents claim comes directly from the subconscious, where they say the "true self" can be found, rather than the conscious mind of the writer. Alternatively, some claim that the writing in fact comes from a source entirely outside of the writer and that the technique allows you to access other intelligences beyond your own, such as spirits and even aliens. Either way the idea is that the writing is not the result of any conscious effort on behalf of the writer, and that is something that I am entirely willing to accept, though whether the messages actually come from your "true self" or from Martians is another matter entirely.

The description of automatic writing seems uncannily similar to that of the ideomotor effect itself. It is after all the unconscious movement of the hand that comes as a result of the expectation of a result. There are numerous websites and books that provide instructions on how to try automatic writing for yourself and here is a list of the points they all seem to agree upon so that you can do just that. As you read through them keep in mind that, whether automatic writing is really powered by spirits or not, they do set things up perfectly for the ideomotor effect to come into play.

1. You will need a pen, or pencil, and several sheets of paper.
2. Find somewhere nice and quiet where you will not be distracted.
3. Sit yourself at a desk or table so that you are in a comfortable writing position.
4. Take time to clear your mind, taking deep breaths so that you feel relaxed.
5. Hold the pen or pencil in a normal fashion and touch it to the paper.
6. Do not try to consciously write anything.
7. Take your time; it may not happen straight away.

8. Keep your mind as clear as you can and just allow you hand to move as it will.
9. Do not look at the paper, in fact you may find that closing your eyes helps.
10. If you feel a break in the flow move to a new line or a new sheet of paper.
11. If you do find that you have produced something look over it carefully. It is likely that the writing will at first glance appear as scribbles or nonsense, but do your best to decipher what is there.
12. You may find that you have produced images or symbols as well as letters and numbers and so keep an eye out for them as well.

And that is pretty much all there is to it. A number of the sites suggest that once you get the hang of it you might like to try asking questions and see if you get answers. They also mention that if you are of a nervous disposition then it might be best to stay clear as you may find some of the answers you get disturbing. Now does that remind you of anything? *Cough* Ouija boards *Cough*. Anyway those instructions seemed easy enough to follow and so I of course decided to give it a go myself, and here is the best of what I came up with:[*]

---

[*] I really believe that I gave automatic writing the best chance I could and yet I still feel I was doing little more than scribbling with my eyes closed. I spent a good hour producing page after page of scrawling marks following the instructions listed above as well as a few alternative methods that I came across. I produced maybe twenty pages doing my best to keep a clear mind and think of nothing. Another dozen were made focusing my thoughts on a departed loved one and around the same number doing my best to "open myself up to what the universe had to say". I also tried the technique of thinking of a letter and trying to start each line with it but otherwise letting things go where they will. This felt rather like cheating to me but I produced probably ten pages like this

Building your Skeptical Toolkit

[Image over page]

---

using the ten most common letters I could think of as my starting point. Then both myself and a third party spent a further forty-five minutes or so going through the pages looking for anything that resembled letters, numbers, words or pictures. I then spent a further hour on my own going through them again and, with the help of the internet, searching for anything that resembled a letter, number or common words found in around a dozen or so languages that do not use our standard English alphabet. These included Arabic script, Egyptian hieroglyphs, Cuneiform, Chinese characters, the Greek and Coptic alphabets as well as that belonging to the Phoenicians. Luckily, I have a tablet and pen setup on my computer and so was able to record my attempts at automatic writing into a format that is readily editable. As such I was then able to take what I believed to be the best examples from the dozens of pages and copy them all together into one page to produce the image used in this book. I also made sure that I worked through the pages in order when doing this, so that the best from page one appear at the top left of the picture followed by the best from page two etc., so that if any consistent message was trying to come through over multiple pages then the order would not be disturbed. And yet despite all of that I remain completely unconvinced that anything other than the ideomotor effect and a good dose of pareidolia, discussed in detail in the next chapter, was at work here. Still you can't say I didn't try.

Your Brain Hates You!

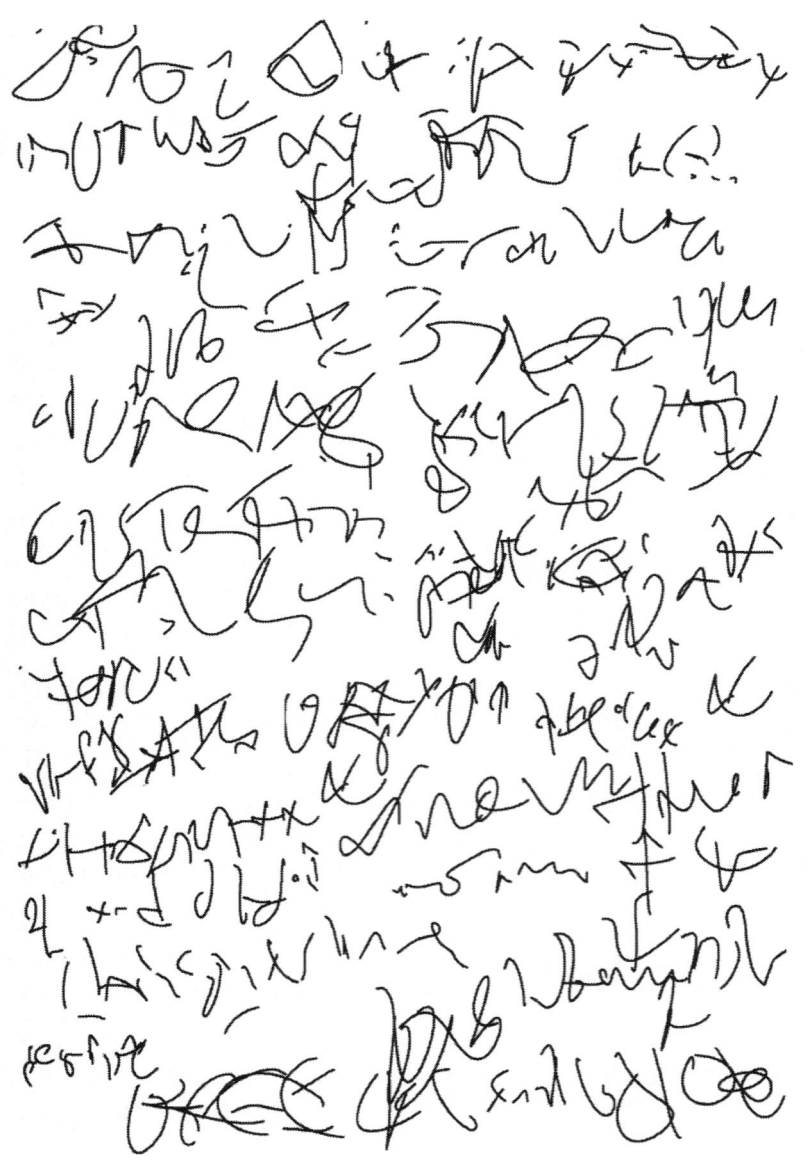

Now I don't know about you but I can see a number of things in there that could be letters, images and even words. All of which brings me nicely onto the subject of the next chapter.

Building your Skeptical Toolkit

Here's looking at you kid
The wonderful world of Pareidolia and patterns that aren't really there

It is a beautiful sunny morning and, as it is the weekend, you've slept late and have awoken feeling hungry. You decide that, for today at least, you are going to say to hell with the diet and healthy eating and treat yourself to a full English breakfast with all the trimmings. You cook the eggs, bacon, sausages, hash browns, mushrooms, beans and tomatoes and pour yourself a tall glass of freshly squeezed orange juice. You are just about to sit down when you realise that you have forgotten an essential ingredient, a few slices of hot buttered toast. You pop the bread in the toaster and get the butter from the fridge. A few minutes later your life changes forever. For as the toast pops up and you drop it onto a waiting plate you find everything you believed about the World called into question. For there, staring up at you from the top slice of toast, is the unmistakable face of Jesus...or the image of the Virgin Mary...or is it the name of Allah? Or maybe, just maybe, something else entirely is going on here.

The human brain is a pattern recognising machine, and this is a task at which it greatly excels. Indeed it could be argued that your brain is a little too enthusiastic in this regard, as it is more than happy to see patterns where none actually exist. In fact, it takes only the vaguest of stimuli for your brain to start applying significance to all manner of random things. For example take a look at this very simple image:

341

Your Brain Hates You!

This image is simply two dots and a curved line and nothing more, and yet I have no doubt that you all immediately identified it as a face. And what's more you all, unless you happen to be a sociopath, will have classed it has a happy, smiling face. And yet of course it is not a face, it lacks probably 99% of the things needed for it to be an actual face. But still, no matter how hard you try, you can't look at those two dots and that curved line and not see a happy, smiling face looking back at you. So what gives?

Psychologists call this tendency for the brain to assign meaning and significance to the most simple and ambiguous of patterns Pareidolia, which, before you start wondering, is pronounced Pa-re-doe-lee-a. And when it comes to vague and barely distinguishable patterns there is nothing your brain loves more than faces.

To the right you will see many examples of this effect in action. That last image makes me feel slightly sad, but then the disbelieving sink in the middle always cheers me up again. So, as before, there really isn't much to these imagines that actually resembles a real face, and yet once again you should have no difficulty picking out the face like shape in each picture, as well as identifying a specific emotion to go along with that face. And what's more, it is very likely that you spotted the faces in those pictures straight away. You didn't need to study the images in order to make out the faces and indeed, as it turns out, you probably identified the faces before working out what the pictures really were.

A study published in the scientific journal Neuroreport showed that the area of our brain believed to be associated with facial recognition, known as the ventral fusiform cortex, picks up an image as a face a fraction of a second before the cognitive part of the brain can interpret it as such. It was also discovered that when it comes to objects that can be mistaken as faces this same part of the brain also kicks in, admittedly a couple of milliseconds slower than with real faces, before our cognitive abilities can identify the object for what it really is. It seems that our brains really do see those pictures above as faces before we see them as alarm clocks, sinks or exhaust pipes.

So why would this be? What possible benefit could there be to seeing faces in the random patterns of everyday objects? Well, while it is still debated as to exactly what is going on, it is generally accepted that these instances of mistaken facial identity are something of an evolutionary by-product related to things that do have a direct benefit to our survival. Carl Sagan put forward the hypothesis that the apparently hard wired ability for people to recognise faces from a young age and from the vaguest of detail, in poor light or at a distance, helped to forge family bonds in our early ancestors, with those better able to recognise faces being more likely to form a connection with their parents.

> "As soon as the infant can see, it recognizes faces, and we now know that this skill is hardwired in our brains. Those infants who a million years ago were unable to recognize a face smiled back less, were less likely to win the hearts of their parents, and less likely to prosper. These days, nearly every infant is quick to identify a human face, and to respond with a goony grin."

On top of this Sagan suggested additional benefits to being able to quickly identify not only faces but emotions in our more hostile evolutionary past. Someone who is able to distinguish friend from foe in a fraction of a second is more likely to avoid unpleasant consequences than someone who is unable to do so. However it seems that the ability to do these things successfully comes with the trade-off of occasionally seeing faces where none exist. In fact, as some suggest, it may be impossible to have one without the other.

Imagine for a moment that you are an early human out for a casual stroll around the African savannah. As you pass an area of long grass you glance over and see something that looks like the

Building your Skeptical Toolkit

face of a lion staring back at you. You immediately turn and run in the other direction as fast as your legs will carry you and thus live to casually stroll another day. Now if it turns out that what you actually saw was not a lion but just a pattern formed by the interaction of light and shadow in the long grass what have you lost, other than a little self-respect that is? The answer is nothing. But what if the situation had been reversed? What if there had really been a lion looking back at you and you failed to recognise it as such? Well we wouldn't be having this conversation for one thing. It would seem that the evolutionary advantages of being able to recognise faces from the smallest of information far outweigh the negative effects of false positives. Those who see faces, and therefore potential danger, all around them are far more likely to survive than those who do not. Thus, we escape being eaten by lions but have to live with seeing faces in hand luggage and bathroom furniture.

However, Dr Takeo Watanabe, a neuroscientist at Boston University, thinks that there may be more to it than this. His research indicates that part of the reason we see faces everywhere may be because, well, we really do see faces all around us most of the time and out brains simply get used to seeing them. In one study, published in the Proceedings of the National Academy of Sciences, Dr Watanabe showed that if the brain is assailed with a specific stimulus it can continue to perceive that stimulus as being present even when it no longer is.

In order to demonstrate this effect Dr Watanabe and his colleagues sat subjects in front of a computer screen that had a series of faint dots moving across it. Initially the subjects were unable to tell in which direction the dots were moving, however in the second set of tests Dr Watanabe superimposed letters over the dots making it easier for the participants to identify in which direction they were traveling. The third part of the test was where

345

things started to get interesting. Here Dr Watanabe presented the subject with a blank screen and again asked them to describe what they saw. Surprisingly the subjects insisted that they could still see dots moving about the screen and also tended to state that they were moving in the same direction as the previous real dots they had seen.

Dr Watanabe believes that these results show that it is possible for our brains to learn something "too well" and that this can actually affect the way in which we see reality. In his experiment the subjects got so used to seeing dots that they continued to see them even when there were no longer any dots on the screen. A similar thing could be happening when it comes to faces. We see faces everywhere and as a result our brains become saturated with them. This, Dr Watanabe concluded, could explain why our brains produce so many false positives.

But why does any of that matter? After all pareidolia is harmless. I mean it is not as though anyone would take a random pattern that just happens to look a bit like a face and claim that this coincidence implies some deep underlying significance and as such should affect the way we live our lives, or would they? Sadly I expect you already know the answer to that question.

On July 25 1976 the Viking Orbiter, part of the Viking 1 spacecraft sent to Mars by NASA, was scouting for a potential landing site for its sister ship, Viking 2, and took a picture of an area of Mars known as the Cydonian mesa, an area of rock almost 2km from end to end. The image was sent back to Earth and when examined the scientists at NASA discovered what appeared to be a human face staring back at them. The original image was dismissed by Viking chief scientist Gerry Soffen as simply a "[trick] of light and shadow" but when a later image appeared to show the same thing it caught the attention of those with an interest in extra-

Building your Skeptical Toolkit

terrestrial intelligence and alien visitation and thus the "Face on Mars" phenomena was born.

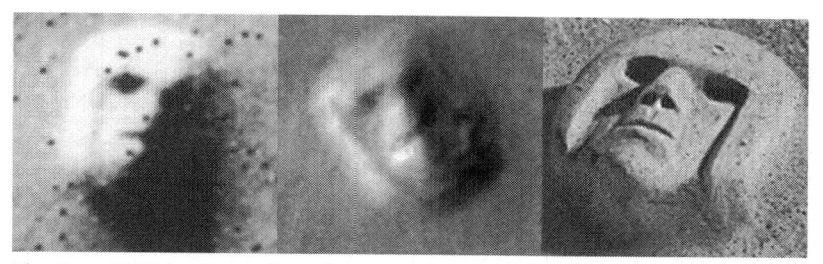

The two 1976 images of the "Face on Mars" and an artist's impression of what it was believed the face looked like. It is important to note that what appears to be a nostril in the first image is actually an error in the data and not part of the face itself.

The main proponent for the idea that this is in fact a real face carved by some ancient Martian civilisation is conspiracy theorist Richard Hoagland, who really brought the face to the public's attention with his book *Monuments of Mars*, in which he claimed that the face was only a small part of a much larger collection of artificial structures that littered the Martian landscape. Hoagland claims that if you study the Viking pictures carefully you can see evidence of entire cities, complete with roads, temples, skyscrapers, power stations and pyramids, hidden beneath the Martian sands. To say that astronomer, writer and former president of the James Randi Educational Foundation, Dr Phil Plait disagrees with him is to put it very lightly.

Plait has examined Hoagland's claims and believes that he is "grossly misinterpreting" the images, as well as falling victim to more than a little pareidolia. He has shown that what Hoagland believes to be buildings hidden beneath the sand of Mars are in fact compression artefacts caused by the way in which Hoagland manipulates the digital images he uses to make his case. I'm going

Your Brain Hates You!

off on a tangent somewhat here but let me quickly explain how this works.

Think about a modern digital camera. Generally a good digital camera will take a picture that is several megapixels, or a million pixels, in size and as a result the file size of the picture itself will be rather large. For example the camera I used to take the original of the picture shown below is a few years old now and yet still produced an image that was 5.2 megapixels and just over 2.5mb in size. In order to reduce the size of a picture file it is possible to use various mathematical compression techniques, one of the most popular being JPEG. The JPEG file system uses an adjustable scale of compression that is known, rather aptly it seems, as "lossy" compression. This refers to the fact that in order to reduce the size of the image the compression method loses some of the data stored in the original image. As I said the level of compression with JPEG files can be adjusted with low levels of compression producing a much smaller file size with almost no visible loss in picture quality (see image A), while at maximum compression the quality of the image is also drastically reduced (see image B).

Things get worse if you combine high compression with magnification of the image. I took the highlighted section of image A, resized it so that it was the same size as the image it was

Building your Skeptical Toolkit

taken from and then saved it at maximum compression. The result is the bizarre array of blocks and shapes shown in image C. Now of course none of those square blocks actually appear on the real flower, nor in the original image for that matter. They were all created by the JPEG compression algorithms.

So let's bring this back to Hoagland's Martian cities. As I previously mentioned Dr Phil Plait believes that the objects that Hoagland says are Martian cities are just the result of images that have been overly magnified and overly compressed. In short those Martian tower blocks and power stations are simply compression noise. None of them are really there.

This also brings us on to an element of pareidolia that I haven't mentioned yet. Earlier in the book we took a look at confirmation bias and the way in which our brains tend to filter information in a way that supports our pre-existing beliefs and attitudes. Well bias plays a similar role when it comes to pareidolia as well. Here our brains tend to be biased with regards to our own experiences, that being things we have seen before, the context the image is seen in and our expectations regarding what we think we should be seeing. It is not hard to see how someone like Hoagland, who really believes the face on Mars is an artificial structure, could look at compressed images of the Martian surface and, in that context, see objects that, admittedly, do vaguely resemble buildings and thus believe there to be a city on Mars. His expectation, having seen what he believes to be one artificial structure, is that there must be more nearby. As such that is just what he sees. As for the experience aspect, well Hoagland regularly uses aerial images of buildings on Earth as comparisons to what he believes to be similar buildings on Mars. His experience of what buildings on this planet look like from above is shaping what he looks for when searching for buildings on Mars.

349

Your Brain Hates You!

But what about the face itself? I mean that is not a compression artefact, it clearly shows up in the original images taken by Viking 1. What's going on there? Well unsurprisingly the answer is the pareidolia effect. Over twenty years after the original pictures had been taken and had caused such a stir NASA sent the Mars Global Surveyor (MGS) back to the Red Planet and on April 5th, 1998 it took a new picture of the Cydonian mesa that was ten times sharper than the original Viking images. When the pictures were uploaded to the JPL (Jet Propulsion Laboratory) website it quickly became clear that the face was not a face at all, but rather a natural landform.

This of course didn't satisfy everyone, with some complaining that the cloudy weather at the time the image was taken had obscured the details of the alien face, and so on a clear summer day in 2001 the MGS, its camera set to maximum resolution, took the clearest picture of the area yet. On top of this a device on board the MGS called the Mars Orbiter Laser Altimeter, or MOLA, took hundreds of altitude measurements of the various mesas in the area, including the face itself, and was thus able to compare them to one another. The results? Well in the words of Jim Garvin, chief scientist for NASA's Mars Exploration Program, when the face was compared to the others mesas in the area it was found that *"all of its dimensions, in fact -- are similar to the other mesas. It's not exotic in any way."*

Building your Skeptical Toolkit

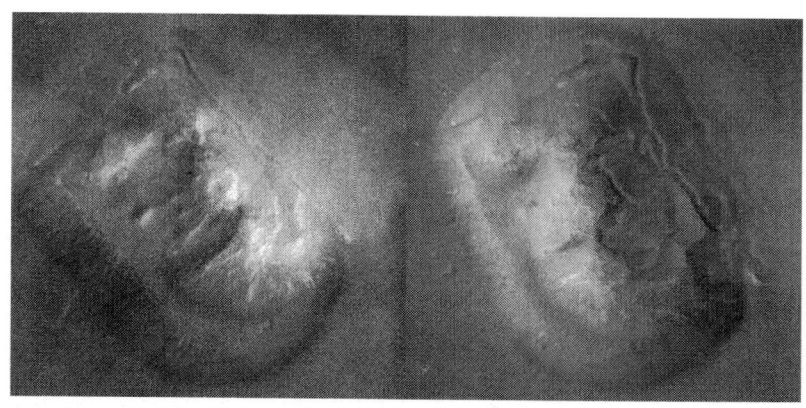

Two images of the Face on Mars taken by the Mars Global Surveyor. The first image was taken in 1998 and the second in 2001. Each pixel in the 2001 image represents just 1.56 meters of the Martian landscape, compared to the 43 meters per pixel which was the best 1976 Viking photo could manage.

But while the Face on Mars is perhaps one of the most recognisable examples of pareidolia in action it is not the most common example you are likely to encounter. For that we need to leave the extra-terrestrial and come a bit closer to home to take a look at the wonderful, and it appears potentially lucrative, world of religious pareidolia.

As we have already seen pareidolia is the brain taking random patterns and shapes and applying undue significance to them. What we see when we look at these patterns is also often influenced by our pre-existing beliefs and expectations. As such it shouldn't come as a surprise that when it comes to pareidolia in the media the stories that come up time and again are those involving people with deeply held religious convictions and images that imply strong spiritual significance.

One famous recorded example of this is the Jesus Tortilla. On October 5th, 1977, Maria Rubio, from the small town of Lake Arthur, New Mexico, was cooking up a batch of fresh tortillas for her husband Eduardo's breakfast. To her surprise she discovered that the skillet burns upon one particular tortilla formed an image that was, to her at least, unmistakably the face of Jesus. She immediately consulted her husband and daughter, Rosy, who confirmed that they too could see the face of Christ staring up at them from the tortilla.

Word travelled fast and by the time young Rosy returned home from school that day there were already people waiting to witness this "miracle" for themselves. *"That afternoon when I drove in from school, there was already a line of people waiting in front of my parents' house to go in and see the tortilla,"* said Rosy Rubio, *"It was amazing."* But it didn't stop there. The Rubio family set up a shrine to the tortilla in their backyard that, until the tortilla was broken in 2006, attracted many thousands of pilgrims from all over the World. Even after it was damaged the Rubios still keep the now somewhat worn tortilla in their home.

And then of course there is the story of the Grilled Cheese Virgin Mary. This story starts out in a similar fashion to that of the Jesus Tortilla but then becomes even more amazing. In 1994 Diane Duyser made herself a grilled cheese sandwich and had taken a bite when she noticed the calm loving eyes of the Virgin Mary looking back at her. *"I went to take a bite out of it, and then I saw this lady looking back at me,"* Duyser explained. She then did the obvious thing. She put the sandwich in a plastic bag and put it into her freezer where she left it for the next nine years.

The story continues in 2003 when Duyser decided to put the sandwich up for auction on eBay. At first eBay took it to be a joke and pulled the sale. However once Duyser provided photographic proof that she would be able to deliver on the bid eBay reinstated the sale of the sandwich where it was quickly snatched up by GoldenPalace.com, an online casino, for the rather tidy sum of $28,000. Yes that's right, twenty eight thousand dollars for a ten year old grilled cheese sandwich with a bite taken out of it. Golden Palace executives said they were willing to spend *"as much as it took"* to own the sandwich. Spokesman Monty Kerr explained to The Miami Herald that *"It's a part of pop culture that's immediately and widely recognizable...We knew right away we wanted to have it."* Now it is unclear if the Golden Palace really believes it to be the face of the Virgin Mary on the sandwich, however Diane Duyser remains convinced, stating that *"I would like all people to know that I do believe that this is the Virgin Mary Mother of God"*.

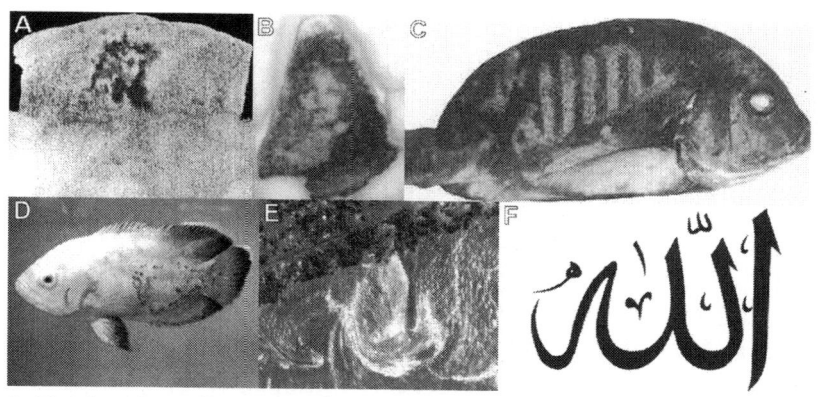

A. The Jesus Tortilla. B. The $28,000 Grilled Cheese Virgin Mary. C & D. Two examples of Allah fish. E. The Sri Lanka tsunami, December 26th 2004. F. The word Allah in Arabic calligraphy for comparison.

But it would be unfair and inaccurate for me to imply that this sort of thing was limited to followers of the Christian religion as similar images appear in association with many of the World's religions. Therefore allow me to introduce you to the Allah Fish. There are many examples of fish who apparently bare the name of Allah in Arabic script upon their bodies but I will just mention two of them. The first was caught in Dakar, Senegal by a Lebanese man by the name of George Wehbi. After bringing his catch home

Your Brain Hates You!

his wife apparently noticed that one of the fish bore natural markings that looked very much like Arabic writing. Wehbi took the fish to Sheikh al-Zein who translated the writing to say "God's Servant" (on its body), "Muhammad" (near its head) and "His Messenger" (on its tail). Now while there are pictures of this fish, see image C in the picture on the previous page, I was unable to locate any further information concerning the details as to where exactly on its body the alleged words are said to be, as from the picture only those on the main part of its body are clear. I guess we will all have to take this one with a healthy dose of salt flavoured skepticism.[*]

Another example for which there is plenty of information however comes in the form of an Oscar fish, with the somewhat unoriginal name of "Oscar", from the town of Waterfoot, Lancashire. In early 2006 locals began flocking to their village pet shop, Water Aquatic, after it was discovered that a recently purchased albino Oscar fish bore markings on one side of its body that closely resembled the Arabic word for Allah.[†] Excitement grew when the other side of the fish was examined and markings said to form the word Muhammad were also found. Within hours, the fish had attracted the attention of not only the local village

---

[*] Why then would I choose to use this example if the information on it is so limited? Well in the end I went with it because it is the most common example of an Allah fish found on Muslim websites. It only seemed fair to use what the believers themselves appear to consider the best example. That said there is some indication that this particular example may be a fake, however I could find nothing definitive on this and so decided to include it as an example of the concept if nothing else.

[†] I feel I should mention that I was forced to edit the picture of this fish slightly in order to get the markings to show up in black and white. The original fish is a creamy-grey in colour with the markings appearing in orange. It seems these colours appear pretty much identical when transferred to grey scale and as such I had to make the markings appear darker than they are in reality.

Building your Skeptical Toolkit

residents but of national print and television news reporters as well.

Once again it seems that people had little trouble believing that the all-powerful Creator of the Universe would choose to communicate with them by writing messages on the side of fish. Mohammed Riaz-Shahid, the 38 year old manager of the Oasis Fast Food restaurant across the road from the pet shop, was one of the leading voices in support of this being a real message from God. *"There's no doubt about it,"* he said. *"The markings are clear to see. Allah on one side, Muhammad on the other...Christians, Jews - they all believe in God...So God's showing a sign to everyone, a sign that he's here."* Others echoed his words, for example Peter Hurst a 17 year old from the village said *"It's a sign of something - no doubt - probably God."*

Of course examples of religious pareidolia are not always found in such innocuous places. On December 26th 2004 an undersea earthquake off the west coast of northern Sumatra, measuring around 9.3 on the Richter scale, sent a tsunami roaring across the Indian Ocean devastating many countries, including Sri Lanka, and resulting in over 230,000 deaths, hundreds of thousands of people being injured and millions rendered homeless. Within days Mohamed Faizeen, manager of the Center for Islamic Studies in Colombo, Sri Lanka, had presented a satellite image taken mere moments after the tsunami smashed into Sri Lanka's west coast as proof that the tsunami had been sent directly by God himself as a punishment for ignoring His laws.

The image, taken by the Digital Globe Quick bird satellite on Dec. 26 at 10 am local time, shows an area of the west coast close to the town of Kalutara shortly after the tsunami impact and as the water was starting to recede. The combination of whorls and side waves do produce a pattern that looks vaguely like that Arabic script for Allah, though I for one feel that you have to work to see this one. Faizeen on the other hand seems far more convinced. *"This clearly spells out the name 'Allah' in Arabic,"* he said, *"Allah signed His name...He sent it as punishment. This comes from ignoring His laws."*

Before we move onto the final example of religious pareidolia that I want to cover here let's take a moment and think back to the end of the previous chapter when we were looking at the subject of automatic writing. As I implied in the closing lines of that chapter not only is automatic writing the result of the ideomotor effect at work but it is also a prime example of pareidolia as well. Let me remind you of the final two instructions explaining how to practice automatic writing.

11. If you do find that you have produced something look over it carefully. It is likely that the writing will at first glance appear as scribbles or nonsense, but do your best to decipher what is there.
12. You may find that you have produced images or symbols as well as letters and numbers and so keep an eye out for them as well.

You are encouraged to search the scribbles for meaning that may not be obviously apparent. When I did this for myself I was under no illusion that there was really a message trying to come through from either my subconscious or the spirit world; however that didn't stop me from trying to make connections between the scribbles that clearly didn't really exist. One example of this, which if you care to look can be found somewhere in the automatic writing image in the previous chapter, involved six marks that to me seemed to clearly spell out the words "the dux". Now this meant nothing to me and despite the relative clarity of the letters, when compared to the other marks that is, I wasn't

Building your Skeptical Toolkit

originally going to include it in my final example page. However I was then informed by the person helping me go through the pages that "dux" was Latin for duke or leader and that "dux bellorum", which means "duke of battle", was a term used to describe King Arthur. Suddenly we both noticed that the mass of scribbles beside these two words, and which previously neither of us had been able to resolve to anything meaningful, appeared to say "Artur", a derivative of the name "Arthur". Beneath this was a sideways figure eight, otherwise known as the infinity symbol∞.

This, we both immediately concluded, clearly referred to the fact that Arthur is also known as "The Once and Future King" and, legend has it, has promised to return in the hour of Britain's greatest need.

Was all this information really there or did we both simply make connections based upon existing knowledge and transfer this information to the vague shapes and patterns on the page? Additionally, though neither of us really believed that there was a message to be found, it was very easy to conclude that if you had one scribble that referred to something, in this case King Arthur, then the other markings on the pages must do so as well. Another thing, which again fits perfectly with the way in which pareidolia plays with our pre-existing beliefs and experiences, was that I noticed many instances where the scribbles reminded me of the names of friends and family, especially that of my much beloved niece, as well as words related to recent events in my life. At the time of producing the automatic writing pages I had recently resigned from a job that for some time I had not enjoyed. As a result this was still very much in the forefront of my mind and as such I was not at all surprised to see marks that resembled, to me at least, words like "job", "free", "next" and "end".

Bringing this back to the name of Allah appearing on the side of fish and in natural disasters it does not surprise me in the least that, having spotted a shape that does vaguely resemble the Arabic script for Allah, people would then see related terms and ideas in even less clear and more vague markings. The idea that a fish, for example, bears the name of Allah on its scales predisposes you to the idea that other words may also be found elsewhere on its fishy body. Suddenly those marks, which moments before were merely random discolorations, become clear messages from a higher power.

But let us leave gods to one side and for our final example of religious pareidolia let's focus on the darker side of spiritual beliefs. In the wake of the terrorist attacks on the World Trade Center on September 11th 2001 a number of images appeared that, it was claimed by many, provided concrete proof that Satan and his demons were ultimately responsible for the horrific events of that day.

Three of the images used to argue that Satan himself was behind the attack on the World Trade Center on September 11th 2001. The first was taken by freelance photographer Mark D Phillips and the second comes from CNN's television coverage. I was unable to locate a source for the third picture and it is possible that it may be one of many fake "smoke demon" pictures to be found on the internet. However after the first two it seemed to be the most commonly offered example and as such I decided to include it.

Building your Skeptical Toolkit

As photographs and news footage flooded the media following the attack, people all over the World started reporting that they could see demonic faces in the smoke, fire and debris rising from the Twin Towers. One commonly cited example was a photograph taken by freelance photographer Mark D Phillips and sold to the associated press. The image, the first example in the pictures over the page, appeared to show an evil looking face, complete with horns, rising from the smoke coming from the South Tower. Many people wondered if the picture had been altered to include Satan's face but Vin Alabiso, an Associated Press vice president and executive photo editor, was quick to confirm that this was not the case. *"AP has a very strict written policy which prohibits the alteration of the content of a photo in any way,"* Alabiso said when questioned regarding the legitimacy of the image. However he also made it clear that he believed the image in the smoke to be a natural phenomenon rather than a supernatural one. *"The smoke in this photo combined with light and shadow has created an image which readers have seen in different ways."*

The second image that fuelled the idea of demonic involvement came from a couple of frames of CNN's footage of United Airlines Flight 175 striking the South Tower. In this image, the second in the picture to the right, a skull like demon face appears to emerge from the eruption of smoke and flame as Flight 175 hits the tower. Once again there is no evidence that the footage was manipulated in any way to include the face. So what does this all mean, why were these faces appearing in the smoke? Well the webmaster over at christianmedia.us was confident that he knew the answer to this question.

> "Demons can feel and experience things like we can. Consider that picture of the demon below [the third image in the picture above] that has its head sticking up like it is on

359

some kind of rollercoaster ride. An act of hatred and violence is a thrill ride for a demon, they not only participate, in so far as influencing someone to commit acts of violence, they get a thrill during the act. Demons knew what was going to happen in New York and they gathered there to jump in at the point of the impact, like a human jumping onto a moving train to have a thrill.

Most acts of violence are not as huge as this one, or last as long, or kill this many people, so this was Disneyland for demons. There was the planes impacting and exploding and people jumping, and then the buildings falling down and killing over 2,000 people in the process. This was a dream come true for demons."

Could this really be the reason for the faces in the smoke? Unsurprisingly a more down to earth explanation was offered by I.J. Karnats, president of the International Association of Arson Investigators, based in Bridgeton, Missouri. He explained that it is not unusual for people to see strange images in smoke clouds. As a fire heats up it draws in cold air and causes unburned debris to swirl through the smoke. This results in the smoke clouds looking thinner in some places and thicker in others. Karnats also added that natural wind currents can add to this effect. On top of this Vladik Kreinovich and Dima Iourinski, members of El Paso Computer Science department at the University of Texas, conducted a geometric analysis on the satanic faces appearing in the smoke and concluded that *"they can naturally be explained by the physics and geometry of fire."*
So while I guess it would be wrong to completely rule out the possibility of demonic intervention it seems that those who know the most about how fire and smoke behave see nothing unusual

Building your Skeptical Toolkit

in the appearance of the smoke coming from the Twin Towers and posit that pareidolia provides a strong explanation for the things people have seen in the pictures from that terrible day.

Your Brain Hates You!

I think I'm hearing things
Backmasking, reverse speech and EVP

In the previous chapter we only looked at one form of pareidolia, that being visual pareidolia. However there is another way in which this phenomena can play tricks on us, and that's auditory pareidolia. Just as visual pareidolia involves our brains taking vague and meaningless images and applying significance to them, auditory pareidolia occurs when our brains take vague, distorted or random sounds and attempt to interpret them as something more recognisable, usually words. Unfortunately the book format doesn't really lend itself to letting me play you examples of auditory pareidolia and as such I will have to do my best to explain the sounds involved to you.[*] One of my favourite examples of this effect could be described as a case of reverse religious auditory pareidolia, with people hearing religious messages that they really didn't want to be there.

In late 2008 toy company Fisher-Price, as well as their parent company Mattel, started to receive complaints from concerned parents regarding their Little Mommy Real Loving Baby Cuddle & Coo Doll. The doll was designed to make the standard cooing, babbling baby noises associated with young children as well as calling out for its *"mama"*, however this was apparently not what people claimed to be hearing. Instead the callers were insisting that the doll was repeating phrases such as "*Satan is king*" and

---

[*] And where possible I will also include a footnote pointing you to a website address where you can hear the sounds for yourself. I know, I'm just too good to you.

"*Islam is the light*".[*] Fox News ran a story headlined "*Parents Outraged Over Baby Doll They Say Mumbles Pro-Islam Message*" in which Oklahoman Gary Rofkahr, who was amongst the first to raise a complaint about the dolls, said that "*There's no markings on the box to indicate there's anything Islamic about this doll*". Well maybe Gary that's because that's not actually what the doll is saying.

Spokesmen for Fisher-Price were adamant that the doll was not pushing a pro-Islamic message, stating that *"The Little Mommy Cuddle 'n Coo dolls feature realistic baby sounds including cooing, giggling, and baby babble with no real sentence structure…The only scripted word the doll says is 'mama'. There is a sound that may resemble something close to the word 'night', 'right', or 'light'…Because the original soundtrack is compressed into a file that can be played through an inexpensive toy speaker, actual sounds may be imprecise or distorted."* This however did not stop some shops in the US from removing the doll from their shelves.

But this isn't the only toy to be accused of saying inappropriate things. In 1998 parents across the United States were enraged when their children's Teletubby Po dolls appeared to be saying *"Faggot, faggot, bite my butt."*[†] Just two years later the Teletubbies were in trouble again when a Californian woman

---

[*] You can hear the doll for yourself by going to http://www.youtube.com/watch?v=JAVGGCar4YY - I have to say that I can't really hear the "Satan is King" message all that well but I have to admit it does sound like it is saying "Islam is the light" at one point. However if you listen to the original source file installed in the dolls then it is harder to hear these words. The original, if it sounds like anything at all, seems to be saying "gone in the light", but that's just how I hear it. Check out the original recording of the doll for yourself by going to - http://www.shareholder.com/mattel/downloads/Lil_Mommie_Coo_Original_10K.mp3

[†] You can check out homophobic Teletubby Po in action at - http://www.youtube.com/watch?v=vj9oCSu9yYk

threatened a lawsuit against the company making the toys, this time due to a Tinky Winky doll that seemed to be telling children "*I got a gun, I got a gun! Run away, run away!*"[*] And then of course there is the Elmo doll that, in 2008, apparently started making death threats against its young owner.[†]

Now unless you believe that the various manufactures of these toys are intentionally making dolls that deliver disturbing messages then it seems likely that what is being heard is not in fact what is being said. This of course is down to auditory pareidolia, people's brains placing significance onto random noises, but there is another element that plays a major role here and which we should take a quick look at before we move on to examine two of the main areas where auditory pareidolia and skepticism collide. That extra element is often referred to as the power of suggestion or, to use the term used by psychologists, priming.

Priming can be thought of as a set-up used to influence you towards a desired outcome. Let me give you an example that you have probably come across at some point in your life. Say the word "silk" out loud ten times as quickly as you can. Done that? Ok now tell me what cows drink. Now the chances are you are all too clever to fall for that little trick; however, I expect some of you might have answered that question with "milk", when of course the answer is actually "water", due to the word "silk" priming you to give that particular answer.

---

[*] This is a rather interesting example as it shows how powerful suggestion is when it comes to these things. When he says that the doll is saying "I got a gun" that is what you hear. But when he points out it is actually saying "again, again" then that is what you hear. Interesting. Check it out for yourself at - http://www.youtube.com/watch?v=Fl7Xui9lzsk – Is it wrong that I really want one of these?

[†] Hear evil Elmo instructing people to "Kill James" at - http://www.youtube.com/watch?v=4dSqXDHfgLc

Your Brain Hates You!

Ok how about this one. As quickly as you can think of an animal; but not a dog. Odds are that the first animal that came to your mind that wasn't a dog was a cat. For numerous reasons people generally associate dogs and cats together. You are either a dog or a cat person; it is raining cats and dogs etc. As such when asked to think of an animal but specifically one that is not a dog your brain jumps immediately to the animal that most people associate closest with dogs, that being cats. The inclusion of the word dog in the original request primed your brain to go in a predictable direction. However, if you thought of an aardvark instead of a cat that just proves you are seriously weird.

There are various different kinds of priming but they all basically work in the same way. You are presented with some form of stimulus, be it auditory or visual, which influences the response you have to a subsequent stimulus. We can see this effect at work with the Muslim friendly Cuddle and Coo Doll. The News on 6, a news channel based in Tulsa, Oklahoma, decided to investigate this story a little further by asking random shoppers what they thought the doll was saying. They found that at first people had trouble understanding the doll as saying anything specific at all, however, once they were told what the doll was allegedly saying they immediately had no trouble identifying its pro-Islamic sentiments. "*I can see it now after you told me. I can understand how it would be a controversy if it didn't have something on it about the other faiths as well,*" said Louis Hull of Tulsa. Fellow shopper Kathryn Christopher echoed her comment. "*Yeah, that's what it sounds like. It sounds like that now that you've said it. Islam is the light. Huh?*"

In the majority of these cases people are unable to hear anything offensive until after they are primed to listen for specific words and phrases. After that they have no problem identifying the offending statements. This is why news reports on stories like this

Building your Skeptical Toolkit

always make sure that they tell you what you are meant to be hearing before they play you the sound in question, usually with subtitles to reinforce what you should be listening for. Without these priming techniques there is a good chance that you won't hear what they want you to hear and as such have no idea what all the fuss is about.*

Priming plays a big part in the two most common forms of auditory pareidolia that a good skeptic like yourself is likely to encounter, those being backmasking and electronic voice phenomena, otherwise known as EVP. Let's tackle them in order shall we.

Backmasking, sometimes referred to as reverse speech, is the idea that secret messages can be hidden in audio material, mainly, but not exclusively music, which can then be revealed by playing the recording backwards. This idea can be broken down into three main sections, with things starting off firmly in Sanityville but quickly heading off for Crazy Town. Let's start with the most rational part of the backmasking trifecta and the one section for which there is legitimate and genuine evidence.

Simply put some recording artists really do hide backwards messages in their songs. This has been done for numerous reasons including simple artistic preference and satirical humour, as well as to get around censors that would otherwise remove certain words or messages from the recordings. And it seems that

---

* You can find two fun examples of priming in action on Youtube. The first is a amusing song by Indian singer Benny Lava - http://www.youtube.com/watch?v=7Wzujy35cWg – then there is the rather rude Fart in the duck song from the TV show "Kabouter Plop" - http://www.youtube.com/watch?v=cKQ5XNjXWzM – Both are foreign songs with English subtitles showing what it sounds like they are saying rather than what they are actually saying. These also illustrate the interesting fact that, no matter what language the words are actually spoken in, our brains do their best to translate the sounds we hear into the language we speak.

this sort of thing is not limited to any particular style of music, with artists ranging from the Beatles to Frank Zappa, Marilyn Manson to Tupac Shakur all getting in on the act at one point or another. One of my personal favourite examples comes from "Weird Al" Yankovic's 1996 album "Bad Hair Day" in which the song "I Remember Larry" includes the backwards message *"Wow, you must have an awful lot of free time on your hands."*

Many of these intentional messages come as a direct response to the second type of backmasking that we are going to look at. Now at first this may sound very similar to what we have just talked about, as it involves backwards messages found in popular songs, but there are a couple of major and important differences. Whereas the first type involves artists purposefully including backwards messages in their recordings, messages that sound backwards when played forwards, this second type is all about ordinary, forward sung lyrics that, when played backwards, sound like they are saying something else. The other major difference is that generally the artists who record the songs have no idea that these backwards messages exist until someone points them out to them.

The proponents of this particular form of backmasking allege that these hidden messages are included as a form of subliminal advertising, usually for non-conservative or irreligious ideas rather than actual products, that, somehow, our brains pick up on and translate to forward messages in our subconscious minds. Examples[*] of this include the Queen song "Another One Bites the Dust", the chorus of which, when played backwards, apparently carries the message that *"it's fun to smoke marijuana"* as well as

---

[*] To keep things simple for those of you wanting to listen to these things for yourself both of the examples mentioned in this section are taken from the pareidolia episode of Shane D. Killian's excellent Youtube series Bogosity - http://www.youtube.com/watch?v=4r0PrkJQU7Y

Building your Skeptical Toolkit

Britney Spears' first song "Hit Me Baby One More Time" in which Britney, who was 16 at the time, is said to encourage us all to *"sleep with me, I'm not too young."*

This is of course where auditory pareidolia, and a little something called phonetic reversal, comes into play. Now you already know that auditory pareidolia is where your brain tries to impose meaning on meaningless sounds, but what the heck is phonetic reversal? Well the word phonetic refers to the individual sounds that make up a word, otherwise known as phonemes. For example the word "phonetic" is made up of the sounds, or phonemes, fuh-net-ik. As such phonetic reversal deals with the sounds these phonemes make when played backwards. Take the word "kiss" for example. When played backwards the phonemes that make up this word make it sound like the word "sick" or "six". As such Yoko Ono's song "Kiss Kiss Kiss", when played backwards, becomes "Six Six Six" and clear evidence, to some anyway, that she is in league with Satan.

Segueing nicely from that to perhaps the most famous example of this form of backmasking in action, the "Paul is dead" phenomena. In 1966 rumours started to circulate alleging that Paul McCartney from the Beatles had died and had been replaced with a look alike. One of the main pieces of "evidence" in support of this crazy idea involved the phonetic reversal of one of the Beatles songs, "Number 9", which was claimed to carry the message *"Turn me on, dead man."*[*] This, combined with a creative

---

[*] Another alleged clue to the death of Paul McCartney can be found at the end of the song "Strawberry Fields Forever" where people claim to be able to hear the words "I buried Paul." Both Paul McCartney, who wasn't actually dead by the way, and John Lennon have stated that the actual words spoken are "Cranberry sauce." However this is a case of simply mishearing words played forward rather than backmasking or phonetic reversal. Still auditory pareidolia though.

interpretation of the cover of the Beatles' Abbey Road album[*], was enough to convince many that Paul was in fact dead.

This phenomenon reached its peak in the 1980s when fundamentalist preachers in America began claiming that, if played backwards, satanic messages could be found on the records of various prominent rock musicians. For example in his 1982 book *Backward Masking Unmasked* minister Jacob Aranza stated that rock groups *"are using backmasking to convey satanic and drug related messages to the subconscious."*

This led to a spate of record-burning protests across the country, as well as anti-backmasking legislation being filed at both state and federal level. In fact in both Arkansas and California bills were successfully passed on this very issue. In 1983 the California bill was introduced specifically to prevent backmasking that they claimed "*can manipulate our behavior without our knowledge or consent and turn us into disciples of the Antichrist*". Meanwhile in Arkansas that same year their similar bill passed, with a unanimous vote in its favour, mandating that records that contained backmasking must carry a warning sticker that read "*Warning: This record contains backward masking which may be perceptible at a subliminal level when the record is played forward.*"

A tragic story that for me best illustrates the extremes to which things went involves the heavy-metal band Judas Priest. Two days before Christmas in 1985, after listening to Judas Priest records, drinking beer and smoking marijuana, James Vance and Ray Belknap shot themselves in the head with a 12-gauge shotgun.

---

[*] Apparently John Lennon being dressed all in white is supposed to make him look like a clergyman; Ringo Starr in a black suit is meant to represent an undertaker; George Harrison in blue jeans represents a gravedigger; and finally the fact that Paul McCartney is not wearing shoes and is walking out of step with the other Beatles is meant to indicate that he is dead. Yeah I still don't get it.

Building your Skeptical Toolkit

Their parents subsequently filed a law suit against the band, demanding $6.2 million in damages, alleging that they had placed subliminal messages in several of their recordings, including the album "Stained Class," and that these messages, which were said to include the phrases "Try Suicide," "Let's Be Dead," and "Do it, Do it," had driven the two men to try and kill themselves.

The band denied the idea that there were any subliminal messages in their music at all. The band's manager, Bill Curbishley, echoed this to the media. *"I don't know what subliminals are, but I do know there's nothing like that in this music,"* he said. *"If we were going to do that, I'd be saying, 'Buy seven copies,' not telling a couple of screwed-up kids to kill themselves."* Ultimately Judge Whitehead ruled in Judas Priest's favour, but the verdict was less than clear cut. While it was ruled that the band were not responsible for the deaths of Raymond Belknap and James Vance, the court did award the prosecution $40,000 and concluded that their album did in fact contain subliminal messages, resulting in reduced record sales and the band being labelled as the "suicide band" and "subliminal criminals". This was despite the fact that the phrases in question were found to be examples of auditory pareidolia and phonetic reversal rather than intentional messages, and that a near-consensus amongst research psychologist supports the conclusion that subliminal messages, of any sort, are capable of little more than fleeting, subtle effects upon our thinking, and most definitely not anything strong enough to encourage someone to take their own life.

But let's move on now and inch ever closer to our arrival in Crazy Town. Now while intentional backmasking is a genuine technique employed by some artists and phonetic reversal is at least a plausible way by which someone with plenty of time on their hands could hide backwards messages in recordings, the third flavour of backmasking, the pseudoscience known as reverse speech, takes things to a whole new level of weird.

Reverse speech was first promoted in 1983 by Australian hypnotherapist David John Oates and he remains its primary advocate to this day. The basic idea is that our subconscious minds influence the words we use when having everyday conversations in such a way that, should our conversations be recorded and played backwards, hidden messages that reveal our true thoughts and feelings can be found in the things we say. Furthermore the subconscious minds of other people are able to pick up on these backwards messages and understand them in a way that we would recognise as instinct or gut feelings.

But wait, there's more. Oates claims that covert speech, his term for reverse speech, comes from the right hemisphere of the brain while overt speech, normal forward speech, comes from the left hemisphere. Additionally, according to Oates, it would appear that not only does the right hemisphere of our brain have the propensity for speaking in metaphors but also, just like George Washington, it cannot tell a lie.

> "On the surface level, it can act as a sort of Truth Detector as Reverse Speech will usually correct the inconsistencies of forward speech. If a lie is spoken forwards, the truth may be communicated in reverse. If pertinent facts are left out of forward speech these may also be spoken in reverse. It can reveal hidden motive and agenda and other conscious thought processes."

And on top of that he claims that the right hemisphere develops its ability to communicate, admittedly in reverse, before the left hemisphere.

> "Reverse Speech research has discovered that children speak backwards before they do forwards. From as early as 4 months of age they are pronouncing simple words in reverse and by the time they reach 13-14 months of age, complex sentences are forming in reverse."

That's all well and good, but what are the practical implications of such a discovery? Well unsurprisingly Oates believes them to be varied and far reaching. First and foremost there are the therapeutic applications. If Oates is right then recording such things as patient interviews and counselling sessions and then playing them back in reverse could give psychologists an inside track to their patients' subconscious. Oates feels that by using his technique reverse speech could be used to evaluate such areas of the subconscious as personality patterns and behavioural agendas, as well as identify hidden memories and experiences and give psychotherapists an insight not only into their patient's mind but into the physical state of the body as well. Additionally, because reverse speech is said to always speak the truth and reveal the speakers true meaning, by examining a counselling session backwards it could reveal suicidal inclinations or evidence of abuse that the patient is either unwilling or unable to speak of directly. But that's just the tip of the iceberg.

> "Other uses for Reverse Speech put forward include employers using it as a[sic] employee selection tool, lawyers in deposition analysis, law enforcement in the investigation of crimes, and reporters for analysing politicians' speeches."

As well as using reverse speech techniques in his role as a business consultant Oates claims to have been involved in a number of different police investigations and to have even been approached by the FBI to consult on sect leader David Koresh during the Waco siege in Texas. He claims that his analysis of recordings of Koresh revealed that he would have come out of the compound had his grandmother been allowed to go speak to him or if the authorities had backed down. Though he does not come out and say it directly the implication from Oates' writing on this issue seems to be that, had either of these things happened, the tragedy at Waco could have been averted.

Now it doesn't take an expert to see the potential for harm that comes with taking Oates' claims seriously. At the low end of the harm spectrum there is the job seeker who is turned down for a job for which he is aptly suited simply because when his job interview is played backwards it sounds like he is saying something that his potential employer finds offensive. That's bad enough but things quickly get worse from there. How about the counselling session that seems to reveal evidence of abuse when in fact no such thing is taking place? Or the suspect in a police enquiry who incriminates themselves, not by what they say but by what they appear to say when the interview is played backwards. Sticking with the legal theme should people be prosecuted based upon what their conversations sound like in reverse, or a witness's testimony be dismissed because their covert speech shows them to be untrustworthy? It seems Oates sees no problem with this as many of the examples on his website come from legal cases. For example his analysis of the OJ Simpson trial apparently revealed phrases such as *"I killed them high,"* *"Slaughtered them,"* and *"I killed the wife."* Now regardless of your feelings about OJ's guilt or innocence the implications should such "evidence" be considered admissible in a court of law are mind blowing.

If that all sounds very reminiscent of facilitated communication, as discussed earlier in the book, then the comparison becomes even more apt when you realise just how much of such analysis comes down to the listeners own subjective opinion. There does not seem to be any definitive way to identify what, if anything, is said in the backwards messages. As such, it all comes down to what the person listening to them *thinks* they hear, and, as we have already seen, this sort of thing is very open to influence from personal expectations, biases and beliefs.

A clear example of this, at least to my mind, can be found in one of the many Bill Clinton reversals found on Oates' website. Here Clinton is heard to say *"I try to articulate my position as clearly as possible"*, which, at least according to Oates, when played backwards reveals that what Clinton is actually saying is *"She's a fun girl to kiss."* I've listened to it a number of times and even knowing what it is meant to say I have trouble hearing it. I guess it sounds somewhat like this is what Clinton is saying, but what exactly is Oates' evidence that this is actually what he said? As far as I can tell it is nothing more than his personal opinion.

On top of this is the element of reverse speech that I skirted over earlier, that being the idea that covert speech tends to come in the form of metaphors that need to be decoded. One example Oates likes to quote involves the backwards message *"My wolf has fallen in the lake"* which it is claimed means that *"the client's motivation and ambition in life are drowning in a sea of emotion"*. Who is to say if this is the correct interpretation of this metaphor or even if it is a metaphor for that matter? After all I guess it is possible that the client actually owns a pet wolf.

But even if you don't hear the same thing as Oates or interpret the metaphors differently he still has an out. He explains that there is a very good reason why people don't always agree with him.

> "They [reverse messages] are very quick and fast and are often hidden in the high tones of speech. For this reason, speech reversals are very easily missed by most researchers."

With arguments like this it is impossible to ever really show that his claims about what is being said are wrong. Thankfully it seems that very few people take his ideas seriously and there is very little support for his claims from academics in the field of linguistics. Most universities and research institutes simply refuse to test his ideas due to a lack of any real theoretical basis behind his claims, that and the fact that many of his ideas are inherently untestable. However some of the things he says are testable and when these claims have been examined Oates has consistently been shown to be wrong. Let's take a look at a few of these shall we.

First up his is claim that the left hemisphere of the brain produces forward speech while the right hemisphere handles reverse speech. Doreen Kimura, a PhD in psychobiology currently based at Simon Fraser University in British Columbia, Canada, discovered way back in 1968 that this is not the case. She found that both forward and backward speech sounds are identified mainly by the left hemisphere and that even normal speech is not restricted to just one hemisphere of the brain. Oates' claims are simply unsupported by neuroscience. Likewise his claims about children developing the ability to speak backwards before they can talk normally goes against everything that we know about language development.

But what about the words themselves? In their 1998 paper, published in the International Journal of Speech Language and the Law, as well as in the pages of The Skeptic, Mark Newbrook, who holds a PhD in linguistics, and his colleague Jane Curtain conducted a simple experiment in which they tested the ability of people to detect examples of reverse speech taken from Oates audiotapes under various conditions. They found that those people who had been told what they were meant to be hearing were much more likely to hear the messages than those who were listening to the recordings blind, if you will excuse the mixed metaphor. This finding echoes the earlier work of Drs. John R. Vokey and J. Don Read from the Department of Psychology and Neuroscience at the University of Lethbridge. Their research, published in the November 1985 issue of American Psychologist, showed that subjects listening to backwards recordings were generally unable to identify the words they were hearing and that when they did the words were a *"function more of active construction on the part of the perceiver than of the existence of the messages themselves."*

All of which means that refuting the claim that reverse speech can be used to detect lies is a somewhat redundant endeavour. However, this does seem to be an inherently testable claim. All you would need to do is record someone stating a number of verifiable pieces of information, some of which are true and some of which are lies and then simply play these recordings in reverse to see if this results in a higher percentage of correctly identified lies than would be expected purely by chance, or by listening to the recordings normally for that matter. Now as far as I can tell no one has actually carried out this research in any official capacity, but then if all the evidence suggests that the phenomenon of reverse speech doesn't actually exist it is rather pointless to test one of the applications of said non-existent phenomenon. But hey, maybe one of you might like to conduct this experiment yourself someday.

Ok, let's now move on to the second area in which auditory pareidolia and skepticism collide, that being the spooky world of Electronic Voice Phenomenon or EVP. For those of you who don't know, EVP is an alleged method by which spirits can communicate with the living by way of tape recorders and other electronic devices. The idea dates all the way back to the 1920s when no less a person than Thomas Edison, in an interview with *Scientific American*, speculated on the possibility of constructing a device with which to communicate with the dead, though he remained skeptical as to whether there would actually be anything to communicate with.

> "I don't claim that our personalities pass on to another existence or sphere. I don't claim anything because I don't know anything about the subject. For that matter, no human being knows. But I do claim that it is possible to construct an apparatus which will be so delicate that if there are personalities in another existence or sphere who wish to get in touch with us in this existence or sphere, the apparatus will at least give them a better opportunity to express themselves than the tilting tables and raps and Ouija boards and mediums and the other crude methods now purported to be the only means of communication."[*]

However, it wasn't until the 1950s and the invention of the reel-to-reel tape recorder that Edison's idea could really be tested, with Swedish painter and film producer Friedrich Jürgenson generally being credited as one of the first to capture examples of EVP on tape. Things really took off though thanks to Latvian

---

[*] Now it is worth noting that Edison did not say that sound recording devices could be used to communicate with the dead, only that an apparatus might be constructed that would be able to "record" messages from ghosts in whatever format they may be transmitted. It seems to have been taken for granted that Edison was talking about traditional recording devices but given what we know about the man it seems unlikely that this would have been the case.

psychologist Konstantin Raudive, who taught at the University of Uppsala in Sweden. Sometimes working in conjunction with Jürgenson, Raudive made over 100,000 recordings that he described as communications with "discarnate people". Because of his work and the publishing in 1968 of his book[*], *Breakthrough: An Amazing Experiment in Electronic Communication with the Dead*, EVP recordings are sometimes referred to as "Raudive Voices". That said, as with many pseudoscientific claims, modern proponents of EVP like to use more scientific sounding terminology, such as instrumental transcommunication (ITC), to describe the way these alleged voices are recorded using technology. But whatever they call it they are all generally talking about the same thing. Voices appearing on recordings that were not there, or at least didn't appear to be there, at the time the recordings were made.

Now you may be wondering how ghosts and spirits, or even aliens and demons for that matter as EVP does not discriminate, are said to be able to place these messages on the tape recorders in the first place and I will be honest and admit that I couldn't really find any clear information on that. Sure there was a lot of talk about "Trans-etheric phenomena", "The Cosmology of Imaginary Space", "The Survival Hypothesis" and "Etheric-Physical Influences" but nothing particularly coherent. There was even mention of chaos theory, fractals and nonlinear mathematics but, to misquote the great Inigo Montoya[†], *"I do not think those words mean what you think they mean"*. In fact the clearest description of what is apparently going on that I could find came from a

---

[*] Translated into English in 1971.
[†] By far the best character in the epic 1987 movie The Princess Bride. Say it with me. "Hello. My name is Inigo Montoya. You killed my father. Prepare to die."

website instructing people on how to make their own EVP recordings.

> "The basic concept behind EVP is that the spirits of those who have passed on through the process of death can utilize the electrical energy in a recording device to implant a message on the media that can be heard when it is played back."

Right then, we will just go with that for the time being and take it as a given that the dead somehow have this ability to make use of electrical energy and instead focus on how EVP recordings are made. Gone are the days of reel-to-reel recorders. For today's EVP investigator the weapon of choice is a portable digital voice recorder, the kind of thing you might dictate a memo into. And they don't even have to be expensive top of the range recorders. Most EVP investigators claim that the cost of the digital recorder is unimportant and that inexpensive recorders work just as well as expensive ones. Now these sorts of devices are very susceptible to Radio Frequency (RF) interference, but I'm getting ahead of myself there. EVP investigators divide EVP recordings into three categories based upon audibility.

- Class A - Voices that are very clear and easily understandable even without headphones. People will generally be in agreement as to what they say.
- Class B - Voices that are fairly loud and clear but generally require headphones to hear correctly. Not everyone will agree upon what is being said.
- Class C - Voices that are very soft and often so distorted that they're indecipherable. Require headphones and

often need amplification and filtering. Others may not even be able to hear them.*

No matter which class a recording falls into it is important to remember that all classes of EVP recordings generally involve voices that were not audible at the time the recordings were made. Also the voices rarely last for more than a few seconds at a time and, with Class C recordings in particular, EVP investigators often have to spend many hours listening to them over and over again in order to discern exactly what is being said in those few seconds. Voices may be in any language or even a mixture of languages and will sometimes answer questions or even address the EVP investigators directly.

One of the most common methods for recording EVPs is to use a digital recorder with an external microphone attached by a long cord in order to reduce the chance of picking up the sounds of the digital recorder itself on the recording. Other methods include recording without an external microphone attached, recording "white noise" from something like a radio tuned between stations and recording using a crystal set, or a diode receiver, plugged into the microphone socket of the digital recorder. Whichever setup you go with the specific technique for capturing the EVP varies from case to case, as Jennifer Lauer, director and founder of the Southern Wisconsin Paranormal Research Group, explains.

> "We'll go out to the location and we'll interview the witnesses and find out what is going on - what they're seeing and hearing. We'll also take equipment readings to make

---

* Shouldn't that tell you something about whether they are really there or not?

sure what they're sensing isn't an electromagnetic field or radio waves.*

We record EVP in two different ways, depending on what type of haunting it seems to be...EVP can be a residual type of energy. It can be a clip that happened at one time and replays itself like a movie. If it's a residual haunting, we let the tape recorder go in the room to see if we pick up anything.

With an intelligent haunting [where they believe that an actual spirit is present], we would ask questions because we know we would get answers ... We sit down in a group of four to six people. We put the tape recorder in a central location between all of us. We proceed to one by one ask a question to whatever is in the room. After our question, we leave about 20 seconds of airtime for the question to be answered, and then the next person will ask a question."

So that's how they are recorded, so what happens next? Do they just play them back and see what they get? Well yes and no. Some EVP messages can be heard clearly just playing the recording back, though most EVP investigators recommend that you listen to the recordings through headphones with the volume set to a fairly high level rather than on regular speakers. However, other recordings require a bit more work before anything recognisable can be heard. Computer software is used to remove background noise and clicking or hiss from the recording. Next the recording may be run through a number of frequency filters designed to boost sounds that fall into the range of frequencies

---

* This opens up a whole other kettle of fish that I really don't want to go into now. For a good breakdown on why this claim should be treated with skepticism check out Brian Dunning's Skeptoid episode "Ghost Hunting Tools of the Trade" - http://skeptoid.com/episodes/4081

associated with the human voice. EVP investigators also use a technique called dither where they actually add additional sounds to the recordings in order to make the recordings more palatable to the ears. Throw in a little bass boost to amplify the mid-range voice frequencies and you are all set to go.

But what do EVP recordings sound like? Well Class A recordings sound like voices on a badly tuned radio or down an old telephone line. But they definitely sound like voices and discernible words can be heard, even if they don't always make much sense. At the other end of the scale Class C recordings sound like, well let's look at a couple shall we. Dr Michael Daniels holds a PhD in psychology and is an Associate Fellow of the British Psychological Society as well as the Editor of the psychology journal *Transpersonal Psychology Review*. In addition to all that he is also a parapsychologist who has studied EVP and has produced a number of his own recordings, some of which can be found on the parapsychology and psychical research website psychicscience.org. These are specifically selected examples of EVP by a well-respected scientist and as such I think we can take it that they are fairly representative of what you can expect from Class C EVP recordings.[*]

To cut a long story short, they suck. Have you ever played the 1980's computer game Centipede?[†] Well they sound rather like that, an 8 bit, vintage, electronic warbling sound. What's more, those examples are taken from the second incarnation of the

---

[*] To listen to them for yourself and make up your own mind go to the Psychic Science website - http://www.psychicscience.org/evp.aspx
[†] If you haven't then my comparison probably means nothing to you. Well you can find out what it sounds like and play the game for yourself here - http://www.atari.com/centipede.

website. The original examples[*] just sound like static and actually hurt my ears to listen to. The words that Daniels claims to hear are, as far as I can tell anyway, simply not there. There is a slight hint of a few tones that sound similar to those used in human speech, but that is really as good as it gets. You really have to want to hear something there and as far as I am concerned all of the examples I have heard of Class C recordings can be adequately explained by auditory pareidolia.

But what about those Class A recordings[†], didn't I say that there were discernible words and voices in those? Well yes I did, so let's finish out our discussion of EVP by looking at the very best examples they have to offer and the possible explanations for them that don't involve voices from the other side communicating by way of "Trans-etheric Influences".

To start with we need to look at something called cross modulation. Have you ever been out for a drive on a nice summer's day and been listening to the radio when all of a suddenly the signal becomes confused and it sounds like you are listening to two channels at once? Well that's basically what cross modulation is, and it doesn't just affect radios. Most of your home appliances are susceptible to some extent or another, even your toaster and coffee maker. In fact, almost any analogue electrical device can pick up errant FM and AM signals unless their components are properly shielded from RF frequencies.

Ah, I hear you say, but don't EVP investigators use digital voice recorders rather than analogue ones? Well yes they do, but for the most part this doesn't help. You see the microphones on the

---

[*] http://web.archive.org/web/20041018012846/http://www.mdani.demon.co.uk/stunt/jun97s1.htm

[†] Numerous examples of Class A recordings can be found on the website for the Association TransCommunication (ATransC), an organisation that studies EVP - http://atransc.org/

Building your Skeptical Toolkit

vast majority of digital voice recorders are analogue devices and thus susceptible to cross modulation if cheaply engineered, and of course that is ignoring the use of external microphones which generally tend to be analogue as well. And there are more problems than this. Faulty programming, fragmenting RAM and poor analogue to digital convertors are all problems common to lower quality digital recording devices and all help to compound the problems caused by cross modulation.

But that's just the cheap recorders right? Surely serious EVP investigators will want to use the very best equipment that they can get their hands on? Stuff that is properly shielded and grounded and of a sufficiently high quality as to rule out the problem of cross modulation and poor sound conversion, right? Well apparently not. In fact the very opposite appears to be true with EVP proponents, as I mentioned earlier, actually recommending cheaper, low quality equipment. Here is a section taken from the website of the Association TransCommunication, one of, if not the, biggest EVP proponent groups in the World.

> "Recording Equipment—Digital voice recorders are recommended for transform EVP. Less expensive models produce more internal noise which is useful for voice formation. High quality units will probably require added background noise. A computer can also be used, but will probably require added noise."

The EVP Research Association UK takes things even further with this useful piece of advice.

> "We have found that digital recorders are most effective for capturing EVP when used on the lowest quality settings. These settings appear to generate a higher degree of

background hiss which is where many of the voices can be found. This is much like using white noise as carrier signal for EVP voices to come through."

Ah yes, white noise. This seems to be a much sort after product of EVP recordings and once you understand a bit more about it it's easy to see why. White noise is basically the auditory equivalent of the snow you get on a TV screen if you don't have it tuned in correctly. It's what most of us would just call static. EVP investigators do numerous things that result in them raising the "noise floor", that's the electrical noise created by all electrical devices, of their recording devices in order to create as much white noise as possible, as it is here that they believe the voices of the departed can be found. The problem is that white noise can be manipulated to pretty much sound like anything.
Think of Peter Frampton and his famous talking guitar effect. Here Frampton takes the sound coming from his guitar and manipulates it with his voice to make it sound like the guitar is talking. Getting a voice out of white noise works in much the same way. By using a focused sweep filter that moves around the sound spectrum on the white noise you can get a very similar effect to that of a wah pedal on a guitar. This allows you to produce open vowel sounds that, just like Frampton's guitar, can sound very much like human speech. And this is of course in addition to the many other techniques used by EVP investigators to boost the sound quality, such as dither, frequency filters and isolation and base enhancement, which we talked about when looking at Class C recordings.
So let's recap shall we. Most EVP investigators use low end digital recorders that are susceptible to picking up signals from AM and FM radios, TVs, baby monitors, mobile phones, walkie-talkies and shortwave transmitters, to name but a few. They purposefully

take steps to try and boost the "noise floor" in order to produce white noise and then run the subsequent recordings through multiple filters and manipulation techniques, sometimes even adding additional sounds to the mix, in order to tease out messages that, by their own admission, sometimes only they can hear. All of which is assuming they didn't just record a real voice, maybe from another room or next door, that they themselves just didn't hear at the time of the recording.

Or the dead really are communicating by way of "Trans-etheric Influences" and chaos theory. You decide for yourselves.

But let's wrap this all up shall we. It has taken two chapters to go over it all, but as you can see, there is a lot more to pareidolia than first meets the eye. I just want to end by very quickly, I promise, touching on the subject of apophenia and ghosts.

Apophenia is very closely related to pareidolia. In fact pareidolia is often described as a form of apophenia. So what the hell is it? Well apophenia is the term used to describe the experience of making connections or seeing patterns in either unconnected or meaningless events or data. This is something we all do from time to time. Common examples of this that people often attribute to supernatural intervention include little things like a specific person or event spontaneously coming up in a number of different conversations. Maybe the Universe is trying to tell you something? Or, as we talked about earlier in the book, finding out someone died the very night you dreamt about them or the phone ringing and it being the very person you were just thinking of calling. Maybe you have psychic powers?

But I wanted to talk about ghosts here and to my mind it is very clear how apophenia, and of course pareidolia, can explain any number of experiences that someone who is already open to the idea of spirits might have. Have you ever been talking about a departed loved one only to hear a door somewhere in the house slam? Could that be the spirit of the deceased trying to make contact? Maybe you've gone to investigate and, opening the now closed door, you see what looks like a figure standing in the room. You quickly turn on the light but the figure has gone, only an old coat hanging from the cupboard door is visible where the figure once stood. Did you just see a ghost?

Taking things to the extreme a paranormal investigator may find themselves in a situation where they have produced an EVP recording, taken a wonderful picture of something that looks decidedly ghostly and experienced an otherworldly shiver down their spine, all at the same time. Does this mean they now have good evidence of ghostly activity, or are they simply making an unfounded connection between various different events all with perfectly natural, and far more likely, explanations?

In short we all make connections that don't exist, assigning meaning where there is none. We see faces and patterns around us all the time and hear voices when really there is nothing there but random noise. We are pattern seeking creatures, we just happen to be a bit too good at it. So when your much loved Gran tells you that she saw a ghost there is no reason to conclude that she is lying or that the old dear is finally losing her mind. There is every reason to accept that she really did see something, something that she couldn't explain and which in all likelihood did resemble a face or figure in some way. But does that mean that what she really saw was a ghost? If I were you I would be skeptical, there are just too many other alternatives.

Building your Skeptical Toolkit

Skeptical Sidebar #8
Is seeing really believing?

Take a look at these two pictures of a church tower and answer this simple question. Were these two pictures taken at the same angle?

Well believe it or not the answer is yes. In fact both pictures are completely identical, with the second simply a copy of the first. But hang on a cotton picking minute, that picture on the right looks as though it is leaning far more than the one on the left, they can't possibly be the same. Well I assure you that they are and that the apparent difference lays not with the pictures themselves but in that grey squishy thing between your ears. But

Your Brain Hates You!

hey, feel free to put a ruler up against the page if you don't believe me.

So what is going on here, why do these two identical pictures look so different? Well it all has to do with the way in which our brains handle perspective in two dimensional images, as well as the fact that our brains can't help but see the two pictures as being part of a single scene.

Normally when we look at a picture of two towers standing side by side and rising at the same angle the outlines of the towers appear to converge as they recede away from us due to perspective. Our brains take this into account and allow us to interpret a two dimensional image as though it were in fact a three dimensional scene. However when we look at a picture of two towers rising up parallel to one another our brains mistakenly assume that this means that the towers are diverging from one another and as a result this is what we appear to see happening. Let me clarify that a bit more using a picture of the Petronas twin towers in Quala Lumpur.

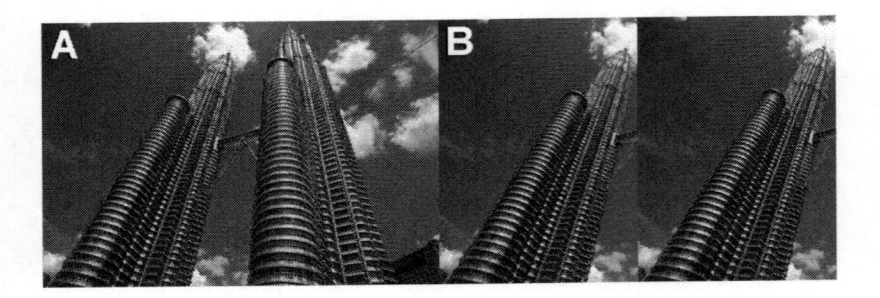

In image A the outlines of the two towers are at different angles and as such the towers converge towards one another due to perspective. Our brains however are able to treat this 2D images as though it were a 3D scene and so create the illusion that the two towers are rising at the same angle, which of course in the real world they do. So when it comes to image B our brains try to

Building your Skeptical Toolkit

do the same thing. In this case the two towers really do rise at the same angle and as such they do not converge together as they do so. Our brains examine the scene and try to interpret it as though it were a 3D image. Now if this were the case the only explanation as to why two objects would not appear to converge as they recede away from view would be if they were in fact diverging from one another. Your brain therefore comes to the natural conclusion that this must be what is happening and as a result this is exactly how things appear to you.

And no matter how hard you try or how clearly you understand what is really going on you can't stop yourself from seeing the two images as diverging from one another. And it doesn't just happen with towers. In fact any scene where objects recede into the distance can be twinned up to create this illusion and it is really easy to make your own.[*]

Now take a look at the picture below. Would you believe me if I told you that all the horizontal lines in this image are parallel to one another and run perfectly straight? Try turning your head, or the book, slightly to one side or the other and see if it changes the way the picture appears. It seems that, after all, seeing really isn't believing.

---

[*] I don't really need to give instructions as I am sure you don't need them, but here they are anyway. Simply take a picture of something straight that recedes into the distance, such as a road or building, so that it appears at a slight angle. Then simply use a photo editing program to put two copies of the picture side by side and your brain will do the rest. Instant optical illusion.

Your Brain Hates You!

Aliens, Witches and Demons, oh my!
A nightmarish look at sleep paralysis

It's night and you are sound asleep in your bed. All at once something wakes you from your slumber and it doesn't take you long to realise that something is very wrong. Though it is too dark to see you can sense that there is...something in the room with you. Something that doesn't belong. Something that causes the hair to stand up on the back of your neck and a chill to run through your body. Something malevolent, unnatural, something...evil. You try to move but find that your body is frozen in place. You open your mouth to cry for help but no sound comes. Fear grips you. Not the weak, everyday fear you have when seeing a spider or when a large dog barks at you, no this is something different. This is a fear that rises up from deep within you and washes over you like a wave of ice cold water. Goosebumps appear on your flesh, sweat breaks out on your forehead and your heart starts to hammer in your chest. No this isn't just fear, this is terror. And now the unseen something is coming closer, you can feel it pressing down upon you, a heavy constant weight on your chest making it difficult to catch your breath. You start to hear strange noises all around you. There are voices in the dark and yet they are unlike any voices you have ever heard before. And now there is light and strange shapes, unearthly images that are both somehow familiar and completely foreign to you at the same time. It seems the stories were true after all; the aliens, demons, witches, goblins or maybe even Satan himself, have finally come to get you. At last a scream bursts from deep inside you and then, all at once, it is over. The lights, the voices and the evil something pressing on your chest all vanish and you are suddenly able to move again. All that remains is the memory...and the fear.

Ok so that is a somewhat extreme example, but has anything like that ever happened to you, because it has to me? How about this? You're in bed and just about to doze off when you clearly and distinctly hear someone call your name; in fact you are so sure of what you heard that you sit up and call back to find out what they want...and then you remember that you are the only one in the house. That has also happened to me. A friend of mine awoke one night to find her face being pressed down hard into the pillow and a heavy weight on her back and neck, a weight that suddenly vanished enabling her to sit up and stare round the empty room in a panic. So what does this all mean? Could this actually be evidence of supernatural forces at work, as some claim, or are these encounters with visitors from other worlds as still others suggest? Or is there a more natural, and ultimately more interesting, explanation for all these things?

Guess what? There is. Sleep paralysis, more often than not referred to as hypnagogia[*] in skeptical circles, refers to a period when, either as they fall asleep or upon waking, a person finds themselves aware but unable to move. Often sleep paralysis is also accompanied by an array of additional symptoms. People may see, hear, feel and even taste and smell things that are not really there. They may feel as if their bodies are levitating, changing in size and shape, growing numb, hot, cold or that they

---

[*] Technically the term hypnagogia, pronounced hip-na-go-ja, only refers to sleep paralysis that happens as you are falling asleep, with the term hypnopomia being used to describe periods of sleep paralysis as you wake up. However many have argued that hypnopomia is a redundant term and instead use hypnagogia to refer to sleep paralysis both when falling asleep and waking up. As stated this is how it is commonly used in skeptical circles as well and is how I will use it here.

are leaving them completely. Perhaps the most common sensation of this type, that generally occurs as people drift off to sleep, is that of falling, often accompanied by a hypnic jerk[*] just before they "hit" the ground. On top of all this many experience intense emotional sensations, commonly fear but also euphoric or even orgasmic sensations as well.

Though this may all sound strange and bizarre sleep paralysis is surprisingly common, with around 30-40% of the population experiencing it at least once in their life, 75% of which experience at least one form of accompanying hallucination and 5% who get the whole shooting shebang of auditory, visual, gustatory, olfactory and tactile hallucinations. If any of these things have ever happened to you then don't worry, you're not going mad as sleep paralysis is rarely linked to psychiatric problems. It is more likely that you are just overly tired, stressed out, that your sleep schedule is messed up or even simply that you were sleeping on your back and, as a result, you are not moving smoothly through the various stages of sleep; and in order to get a better understanding of what exactly is going during sleep paralysis we really need to take a quick look at these stages.

There are two main types of sleep, Rapid Eye Movement sleep (REM sleep) and Non-Rapid Eye Movement sleep (NREM sleep). NREM sleep is further divided into four stages, imaginatively called stage 1, 2, 3 and 4[†], and in an average night's sleep we move back and forth between these four stages and REM sleep. Stage 1 is light sleep, as you transition between wakefulness and

---

[*] That little jump you sometimes get when falling asleep that causes you to wake up again for a moment.

[†] The American Academy of Sleep Medicine (AASM) conducted a review in 2008 of the way in which the stages of sleep are divided, resulting in stages 3 and 4 being combined into a single delta or deep sleep stage. However, for clarity I will use the original 4 stages in this description.

sleep. At this stage people can be easily woken and may not even believe they have been asleep at all if asked. Your body starts to slow down and relax, though you may experience the odd muscle twitch or hypnic jerk and retain some awareness of your external environment. Some may also experience hypnagogic hallucinations at this stage. This stage of sleep generally lasts around 5-10 minutes during which your brain's alpha waves, those most common when awake, transition to slower theta waves.

Stage 2 is where an average adult spends most of their time whilst asleep, around 50% of the total night's sleep. At this stage your brain waves are generally slow but with occasional and rhythmic bursts of rapid brain activity known as sleep spindles. Your eyes stop moving, your heart rate slows, your body temperature decreases and so does your muscle activity. If you are prone to talking in your sleep you may start mumbling during this stage.

Stage 3 is where you start to transition between light and deep sleep. Your brain starts to produce deep slow delta waves, though these are still interspersed with smaller, faster brain waves. At this point in the sleep cycle it is often difficult to wake people up, and if you manage to do so they will likely feel groggy and disorientated for a good few minutes.

At Stage 4 your brain is producing those deep slow delta waves almost exclusively. This lasts around 30 minutes and is important to letting you feel like you have had a good night's sleep. While muscle activity at this point is generally low it is during this stage that things like sleep walking, night terrors and bed wetting are most likely to occur. And again if you talk in your sleep you are likely to really get going at this point.

After about 30 minutes of Stage 4 sleep your body is ready to enter REM sleep and it does so by bringing you out of deep sleep and quickly moving back through the previous sleep stages. When

Building your Skeptical Toolkit

in REM sleep your brain is almost as active as it is when you are awake and, as the name implies, your eyes move around rapidly. Your heart rate increases, your blood pressure rises and your breathing becomes fast and shallow.[*] However your muscles become paralysed and for good reason, because it is during REM sleep that you dream and it is only the fact that you are unable to move that stops you from acting out your night-time flights of fancy. In 1959 French physiologist Michael Jouvet conducted a number of experiments on cats with brain lesions that prevented their muscles becoming fully paralysed during REM sleep. He found that during REM sleep the cats would act our complex muscles behaviours, such as running round their cages and engaging in hunting behaviour. If not for our brains paralysing our bodies we too might end up chasing mice in our sleep.

During an average night's sleep you will cycle back and forth through the various stages of sleep, each cycle lasting around 90-100 minutes, spending around 20% of your time in REM sleep. You may even wake up a number of times during the night, though you probably won't remember doing so come the morning. As the night goes on you spend more time in REM sleep and less in deep, stage 4 sleep. The hypnogram on the next page shows a visual representation of how your brain jumps around between these different stages during the night. It's enough to make you feel sleepy.

---

[*] Fun Fact: During REM sleep it is not just our brains and heart rates that get excited. Our genitals also get in on the act with men gaining erections and women experiencing increased vaginal lubrication. This happens regardless of what we may be dreaming about. So gents should you awake with a touch of morning glory after dreaming about your new shed or football there is no reason to be concerned that you are developing a strange fetish.

397

Your Brain Hates You!

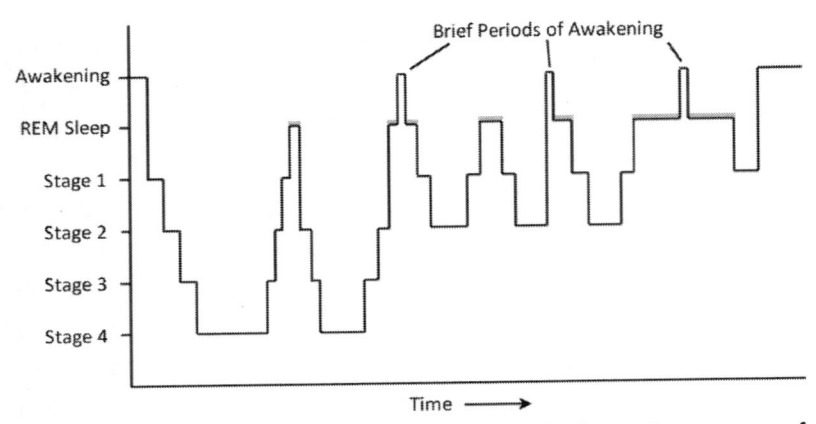

Hypnogram showing the way you move through the various stages of sleep in a typical night.

So now that you have a better idea of what's involved in sleep let's turn our attention back to sleep paralysis and those things that go bump in the night that we mentioned at the start of the chapter. Sleep paralysis tends to occur at two main points in the sleep cycle. The first is during stage 1, as you are starting to fall asleep. As you drift off your muscles start to relax, your body becomes heavy and it becomes harder for you to move or speak. Usually your awareness also starts to shut down during this stage as well, but if this doesn't happen or something causes you to become aware again you may find yourself awake but with your body still nodding off and unable to move. The other time sleep paralysis occurs is at the other end of the sleep cycle, when you are in REM sleep. As you can see from the hypnogram when in REM sleep your brain is actually very close to an awakened state and it is not uncommon for people to wake up for brief periods of time during this stage. Usually this is not an issue and you just drift back to your dreams. However sometimes you may become aware while your body and brain are still in the REM sleep stage and, as your brain has disabled your muscles to stop you acting

Building your Skeptical Toolkit

out your dreams, you can once again find yourself unable to move. Throw in the fact that, even though you are aware, you are still actually dreaming and this can make for a very bizarre and disconcerting experience indeed.

But why should these waking dreams freak even the most rational and skeptically minded of us out so much? Well it is likely that this has a lot to do with something else that happens to your brain when you are in REM sleep. Have you ever had a dream in which you and a giant pink rabbit open a sweet shop on the moon? I have and I can tell you that at the time this seemed like a perfectly normal and everyday thing to be doing. Of course now I can look back and see what a strange[*] idea this was, so why did it all seem to make sense at the time? Well while some areas of your brain are highly active during REM sleep there is one area that is all but shut down at this time, and this is the dorsolateral prefrontal cortex. But why should this matter? Well the dorsolateral prefrontal cortex is the part of your brain that deals with thinking, reasoning and other executive functions. Sometimes described as a "rationality filter" this is the part of our brain that tells us that giant pink rabbits aren't real, and this shuts down during REM sleep and tend to remain inactive during sleep paralysis, thus making any hallucinations we may have seem all the more real.

Ok so let's talk about these hallucinations for a bit. Over the years numerous surveys have been conducted recording the experiences of people who have had sleep paralysis related hallucinations. The results from these surveys show that sleep paralysis hallucinations come in three main types that, for once, actually have interesting sounding names. You have the Intruder, the Incubus and Unusual Bodily Experiences or UBEs. Let's take these in order shall we?

---

[*] But still awesome

Your Brain Hates You!

The Intruder refers to one of the most common forms of hallucination associated with sleep paralysis; that being the feeling that someone or something is present with you in the room when in fact there is nothing there. This sense of a presence can range from a slight uneasy feeling that you are not alone through to complete conviction that something is there with you, and in over two thirds of cases this presence is accompanied by a feeling of abject terror. Now it is more than likely that you have not experienced this sort of sleep paralysis related hallucination yourself, but it is also more than likely that you have had a similar experience during your waking hours. Maybe you have been alone in the house at night and have popped downstairs to grab something. You don't turn on the light as you know the place well enough to find what you are looking for even in the dark. You grab whatever it is and turn to head back to your room when you suddenly get the feeling that you are not alone. At first you are confused, after all you didn't hear anything and you certainly didn't see anything that would indicate that someone else was there with you. It is just a feeling, a sense that someone is watching you from the dark. And then there is the fear. Fear that grips you somewhere deep inside and makes you want to run up the stairs back to the safety of your well lit room. You know it is silly, you know that you are alone and that it is only your mind playing tricks on you; and yet you still take the stairs two at a time and close your bedroom door tightly behind you. Or perhaps you have been walking home late at night and have had the feeling that you are being followed. You look but can't see anyone and yet the feeling of unseen eyes upon you remains, as does the fear that accompanies this sensation. Throw in the inability to move and this describes fairly accurately the basic intruder hallucinations experienced during sleep paralysis.

In many cases this presence is just that, a presence in the room with you, something that is there but which does not interact with you in any way. Just as common is the sense that this presence is watching or monitoring you in some way as you sleep. Some people go further and describe this feeling of being observed as being "stared at" and often attribute malevolent intent to this unseen watcher. Despite my colourful introduction to this chapter people often make much more realistic and every day interpretations as to what this presence may be, rather than jumping straight to ghosts and goblins. Many people will initially attribute this presence to their roommate or spouse, the dog having got into the room or, at the scarier end of the spectrum, a burglar, rapist or even terrorist who has broken into the house. For many it is only after they have realised that their roommate/spouse is out of town, the dog is outside in his kennel and that this burglar/rapist/terrorist appears to be both invisible and completely silent that their minds turn to more supernatural explanations such as demons and alien visitation.

Regardless of what people interpret this presence to be it is very common for people to attribute evil intent to this unseen visitor, sensing that it is something to be feared. In fact it seems that the presence and a sense of fear are closely linked one to the other, though sleep researchers remain unsure if the presence causes the fear or if the fear comes first and is subsequently assumed to be caused by some unobserved presence. Either way the fear is very real and many people have described it as eclipsing anything that they have experienced in their waking lives, not just fear but primal terror. This fear is often accompanied by a sense of doom and the feeling that their life, or more abstractly their soul, is in danger of being consumed or snuffed out by this malevolent evil, which somehow they know is far stronger and more powerful than they are. As such it is not uncommon for people to make the jump to an unnatural cause being behind this overwhelming fear. When I had a hypnagogic episode myself as a teenager I knew, beyond a shadow of a doubt, that there was a demon watching me from the foot of my bed and that if I did not do something it would not only kill me but would drag my soul to hell. The fear I felt in those few seconds as I stared helplessly into the dark trying to see what was there, but at the same time praying I wouldn't see anything, was more powerful than any fear I had experienced before or since. In my mind it was impossible for such terror to have a natural cause and I continued to believe that I'd had a supernatural encounter for many years afterwards, with this belief only really going away when I finally learnt about this bizarre phenomenon of sleep paralysis. The exact nature of the "being" with which people believe they have had an encounter varies wildly, ranging from various demons and monsters to aliens, witches and even man-eating plants, and there seems to be a heavy cultural influence upon what people report experiencing. But we will look more closely at this aspect of the hallucinations in a few pages time.

As well as just sensing a presence there are other experiences that fall under the heading of the intruder hallucinations. Both auditory and visual hallucinations are often associated with intruder hallucinations, though auditory hallucinations are far more common than visual ones. The auditory hallucinations experienced during sleep paralysis come in a wide variety of types and I won't go into detail of all of them here. One thing that does seem fairly consistent however is that these sounds are viewed as external noises that are picked up by the ears rather than as sounds heard only inside your head.

The type of sounds people report hearing may be elementary in nature; buzzing, humming, roaring, scratching, grinding noises. Sometimes these types of noises have a bodily experience to them as well, for example a vibrating noise may be accompanied by the feeling that your body itself is vibrating. Technological sounds are also common, things like sirens, ringing telephones, breaking glass, power tools and other mechanical noises. Natural sounds are equally likely, such as gusts of wind or the sound of waves. Animal noises are also often reported. Some people report hearing sounds of movement such as someone walking around the room or of something being dragged across the floor. Perhaps the most interesting noises reported are those described as human sounds. These can be as simple as moans, laughs and screams but can range in complexity to include things like muffled conversations, distant music and clear but meaningless voices, as well as distinct, clear and understandable words and phrases, more often than not the name of the person having the experience. As I said before, this latter type of auditory hallucination has happened to me on a number of occasions and has been so convincing that I have answered back, only to remember that I was alone in the house.

There are also visual hallucinations, though as mentioned these are far less common and also generally far less vivid as well. Most visual hallucinations tend to be undefined and indistinct, more flickers out of the corner of the eye and implied shapes and patterns than anything clear and definitive. In many cases it seems that what people think they see is strongly influenced by their beliefs and interpretations surrounding the whole sleep paralysis experience. If you sense a presence, feel the overwhelming feeling of fear that accompanies it and therefore assume that you are being watched by a demon, well, it is only natural that when you see a vague and indistinct shape moving towards you that you would see that shape as demonic in nature. And see is the right word here as, while these visions are created by the mind and are far less vivid than normal perception, the overwhelming number of people who have them report them as being fully external events and not images inside their heads. As mentioned the vast majority of visual hallucinations experienced during sleep paralysis are insubstantial and nebulous images, however some people do get full blown, Technicolor hallucinations that can seem just as real and solid as anything experienced during their waking lives. Either way most people report that during the hallucinations themselves, regardless of their strength or clarity, they found the experience entirely compelling, only later during the light of day did they look at them with a more skeptical eye; at least in most cases. As discussed this could have a lot to do with their "rationality filter" being inactive at the time of the hallucination.

Ok let's move on to the second major category of sleep paralysis related hallucinations, the Incubus. This is another common sleep paralysis experience and perhaps the one most closely related with the phenomenon. The Incubus refers to the feeling that a heavy weight is pressing down upon you, usually upon the chest or torso, as sleep paralysis is strongly associated with sleeping on your back, but also sometimes on the back, sides or even individual limbs. The term Incubus is actually a version of the Latin verb *incubare* which means "to lay upon" and describes a mythological creature, a male demon that sneaks in during the night and lies upon sleeping people, usually women and usually for the purpose of intercourse. Stories of creatures like the Incubus, and its female equivalent the Succubus, appear in many cultures, though the version with which most people in western cultures are familiar, whether they realise it or not, is the old English version where, in Anglo-Saxon, such creatures were referred to as Mares. And a Mare that visits you during the night is, of course, a nightmare. Who said etymology isn't fun?

Building your Skeptical Toolkit

Generally the Incubus experience is described simply as the feeling of pressure on the chest or torso; however in some cases it can be more intense than this. People have reported feeling as though they were being crushed by a great weight, pushed or pulled through their bed, smothered, choked or even bitten by some unseen force. As I mentioned at the beginning of the chapter a friend of mine told me of an experience she had in which she felt as though her face was being forced down into her pillow by a heavy weight upon her neck and back making it hard for her to breathe. This is a classic Incubus experience, the difficulty breathing being another commonly reported symptom.

Now obviously there isn't really something pressing down on you so what is it about the sleep paralysis experience that explains this form of hallucination? Well, while the details are somewhat involved the basic idea here is very simple. As we have already covered, when in REM sleep your muscles become paralysed and this includes the muscles used to breathe. In REM sleep your breathing becomes fast but shallow, lots of little breaths that never really make your chest expand that much. Should you become aware during this time it is fairly instinctive to try and take a deeper breath, especially if you are experiencing fear as a result of being unable to move or due to sensing some evil presence in the room with you. The problem is that of course your chest muscles are still paralysed and unable to expand fully. The same goes for the rest of your muscles so that when you try to lift an arm or sit up you find you are unable to do so. Usually during your waking life if you find yourself unable to take a deep breath or lift your arm it is because something is stopping you from doing so, some force is holding you in place or pressing down on you. During a sleep paralysis experience your brain makes this logical leap and assumes, incorrectly in this case, that the reason you can't take a deep breath or move your muscles must be because some force is acting against you and as such interprets your inability to sit up as a weight pressing down upon you and your difficulty breathing as something choking or smothering you.

If this perceived weight or force is experienced alongside the sense of an evil presence in the room that comes with an intruder hallucination it is not uncommon, or all that illogical for that matter, for people to assume that the presence is the cause of the weight on their chest and their inability to move. People combine these two forms of hallucination and conclude that they are being attacked, that something is sitting on their chest or trying to crush them. Once again this can cause intense fear though, unlike with the intruder hallucinations alone where this fear is often nameless and difficult to define, when it comes to incubus hallucinations this fear is almost always related to thoughts of impending death, most commonly dying as a result of suffocation.

The last type of hallucinations often experienced during sleep paralysis events are the Unusual Bodily Experiences (UBEs). These come in two main categories, floating or flying experiences and out-of-body experiences (OBE). Strange as it may sound, given what we have just covered, some people experience the feeling of flying or floating during sleep paralysis episodes. These are usually more gentle experiences, with people often describing the sensation of gravity fading away, of their bodies becoming light, of gently drifting up from their beds or of being lifted and carried. However, as with most experiences during sleep paralysis, there are those who experience the other end of the scale and report rapid and violent movements more akin to a rollercoaster ride or free fall. In the vast majority of cases this is a full body experience, though some do report more localised feelings of weightlessness. Due to the fact that people are unable to move during sleep paralysis these floating sensations are very much a passive experience with the person feeling that they have little control over their movements. That said this lack of control does not itself seem to cause that much alarm and these floating/flying experiences are one of the few sleep paralysis hallucinations that are described as pleasurable events. However as before these hallucinations can be experienced in conjunction with other types, such as sensing an evil presence in the room, and in these cases people once again report feelings of fear, generally related to the feeling that something or someone is taking them somewhere against their will. Can you say alien abductions?

Some people experience these floating sensations in a somewhat different way. For them it feels as if they are leaving their bodies behind and people often report both a feeling of separation from their physical form and observing their bodies from an outside perspective and, often, from an elevated position. In most cases these out-of-body experiences are pleasant events, often associated with positive emotions, though of course there are exceptions to this with people reporting feeling as though they were dying or having their self, their soul, forcibly removed from their bodies. As always these OBEs can happen in conjunction with the various other forms of hallucination and it seems it may be the interactions between the hallucinations that govern whether it is a positive or negative experience.

As we have seen sleep paralysis can have a very profound effect and the hallucinations that come along with it can be highly convincing, due in no small part to the rational part of your brain being shut down at the time. As such it is not at all surprising that over the millennia people have been experiencing these events they have come up with all manner of supernatural explanations for them. These explanations become part of the local folklore and superstitions within the society and as such when someone experiences a sleep paralysis event for themselves they are far more likely to interpret it in light of this existing explanation than to correctly identify what is happening to them. As I mentioned earlier when I experienced sleep paralysis and an accompanying intruder hallucination as a teenager I was convinced that I had been visited by a demon, and remained so for many years afterwards. This conclusion was influenced in no small part by the fact that, at the time, I was strongly religious and already believed in the existence of demons. Throw in the fact that I was a big fan of the Frank Peretti novel This Present Darkness, in which both angels and demons feature prominently, then it is understandable that when, in the dark of the night, I encountered a powerful presence that filled me with terror my mind jumped to the answer most readily available to me, a demon.

Demons are just one of the wide variety of explanations people have come up with, and if you live in Turkey, Greece or Pakistan it is the one you are most likely to turn to. If you are from China or Mexico however you will probably attribute it to ghosts. In areas of Africa the experience is described as "the witch riding your back", while in New Guinea the blame falls on magic trees that feed on people whilst they sleep, and in Iceland goblins are held accountable. We have already talked about the Incubus[*], Succubus and Mares of ancient western folklore and we can add witches or "The Old Hag" to the list as well. The Hag is one of the most enduring and wide spread supernatural explanations and remains popular to this day in places like Newfoundland and some southern states of America such as South Carolina and Georgia.

---

[*] While researching this topic I was surprised and a little alarmed to find that belief in incubus demons still seems to be going strong in our modern society. I discovered people seeking advice on how to deal with attacks from an incubus on Yahoo Answers, specifically emphasising the seriousness of their request, followed by numerous comments from people offering advice on how to go about riding yourself of demonic visitations. Not once did anyone mention sleep paralysis as an explanation.

Building your Skeptical Toolkit

Regardless of who is believed to be responsible for this phenomena the important thing is that throughout history these events have been attributed to specific creatures and beings, all of whom tend to stay fairly exclusively within the culture that first dreamed them up.

A great way to see this sort of cultural influence in action is to look at the supernatural explanation with which we in the west are most familiar these days and that I want to finish this chapter talking about; alien abduction. Up until the 1960s sleep paralysis hallucinations in the USA were generally attributed to the Old Hag explanation, but then in 1965 this started to change and a new player began to take centre stage as the story of the Betty and Barney Hill Alien Abduction hit the headlines. Almost overnight the "Greys" started muscling in on the Old Hags territory as the image of little grey men with big heads pinning people to their beds with cosmic rays started to flood the public's consciousness. Looking at this event we are presented with a few possible explanations for what was going on. It is possible that aliens were turning up and taking over the role previously held by the Old Hag and her witchy friends, or the Old Hag was simply an alien in disguise and the Greys decided it was time to reveal their true identity. Or maybe the cultural zeitgeist changed and so did the hallucinations?

Before we take a closer look at alien abductions I want to make one thing clear. People who believe that they have been abducted by aliens are not idiots, nor are they more suggestible or more given to flights of fancy. When psychologists have looked into this area they have found that those people who have experienced what they believe to have been an alien abduction are no less intelligent, no more prone to fantasy and no easier to hypnotise than those people who have not had this experience. Likewise self-proclaimed abductees are no more likely to suffer from

Your Brain Hates You!

psychopathological conditions than other members of the general public.

A slight reimagining of The Nightmare *by Henry Fuseli that takes into account recent cultural changes.*

These people are not the tinfoil hat wearing crazies you see in the movies; they are just normal everyday people like you and me. That said however the research also shows that "abductees" are significantly more likely to suffer from sleep paralysis and to a greater intensity than people who do not report abduction experiences. They are also more likely to generate false memories and make false associations between unrelated things than non-abductees. And, unsurprisingly, they are also much more likely to hold strong beliefs in the existence of aliens and alien visitation.

With this in mind, is it fair to conclude that alien abductions are probably just the latest in a long line of other worldly explanations for a well-documented and well-understood sleep disorder? Well maybe we should look at what the proponents of alien abductions themselves believe are the signs that someone has had a visit from little grey men. In 1991 the Roper Organization, a public relations and consumer research centre, carried out a poll on behalf of UFO investigators Dr David Jacobs, Budd Hopkins and Robert Bigelow with the aim of estimating how many people in America may have been abducted by aliens. Researchers visited randomly selected adults at their homes and asked a series of questions about their experiences and politics, of which the abduction questions formed a part, and gained replies from 5,947 individuals. In the relevant section people were asked to state if they'd had any of eleven experiences never, once or twice or more than twice. Of these experiences, that the authors believed to be indicative of alien abductions, five were viewed as important "indicator experiences" and it was concluded that for anyone who answered "yes" to four out of five of these there was a "strong possibility" that they were an abductee. 119 people (around 2%) of the people questioned matched this criteria which, when expanded to the population from which these people were drawn, led the authors to conclude that around 4 million people in the US had been abducted by aliens.

So what were these signs, most importantly these "indicator experiences", which led them to this staggering conclusion? Well here they are, along with the percentage of people who reported having said experience, listed with the five important "indicator experiences" at the top.[*]

---

[*] These are taken from Dr David Jacobs 1999 book *The Threat*. Those of you with sharp eyes, and the ability to count, will notice there are only 10 items listed here. The eleventh item on the questionnaire asked people if the word "trondant" had special significance for them. This was a test question, the word "trondant" having being invented by Hopkins to help rule out people who may have felt the need to answer positively to any question asked of them, and as such is irrelevant to the issue at hand.

- 18% had awoken paralyzed and sensed a strange presence in the room.
- 13% had experienced a period of missing time of more than an hour or so.
- 10% had feelings of flying through the air without knowing why.
- 8% had seen unusual lights in their room with no identifiable cause or source.
- 8% had discovered puzzling scars or marks on their bodies that neither they nor other people remember them getting.
- 15% had seen a terrifying figure in the room with them.
- 14% had left their body, or had an unexpected out-of-body experience.
- 11% had seen or sensed a ghost.
- 7% had seen a UFO.
- 5% had dreams about UFOs.

Of the five "indicator experiences" three are readily explained as common symptoms of sleep paralysis, loss of time being a symptom of sleep itself, with terrifying or ghostly figures and out-of-body experiences also fitting perfectly with well documented hypnagogic hallucinations. As for discovering scars that you don't remember getting, well this very morning in the shower I found a bruise on my leg, a cut on the back of my hand and a graze on my shoulder, none of which I have any memory of getting. Does this mean that it is impossible for people to have acquired such things by way of alien abduction, well no, but seeing I have no memory of being probed last night either the possibility that people knock themselves without knowing it or even while asleep should not be ruled out. As for seeing or dreaming about UFOs I will just ask you to remember what the U in UFO stands for; personally I have seen and dreamt of many things that I would describe as "unidentified".[*]

Now while the evidence, even that coming from abduction proponents themselves and taking into account my slightly sarcastic analysis, does seem to strongly support sleep paralysis as the explanation for alien abduction experiences, Dr Jacobs would be the first to remind us that we should not simply dismiss all abduction claims as hypnagogic hallucinations, and he is completely right. As a skeptic it is important that you examine each claim as they come, look at the evidence and evaluate what you are being told based upon what it tells you, rather than upon any pre-existing beliefs you may have. And if we do this I think that the vast majority of abduction stories can be chalked up to good old fashion sleep paralysis.

Let me wrap this up with an interesting and perhaps telling bit of minutiae. At the beginning of the chapter you may remember that I mentioned that around 40% of the population experience sleep paralysis at some point in their lives, with 5% of that number experiencing the full spectrum of hypnagogic hallucinations. You'll also remember that in their poll Dr Jacobs and his colleagues concluded that 2% of the population had probably been abducted by aliens based, amongst other things, on them having experienced paralysis, the sensation of a presence, the feeling of flying and having seen strange lights. Well it may be meaningless and a coincidence but another way of saying 5% of 40% of the population is to say, yup, 2% of the population. Funny that.

---

[*] If you wish to look at or even take an updated version of this questionnaire yourself you can do so on the website of the International Center for Abduction Research, of which Dr David Jacobs is the director. It includes the addition of such definitive questions like "Have you ever dreamed about lying on a table?" and "If you are a woman, have you ever had any unusual problems with pregnancy?" Surely there can be no other explanation for these things than aliens? - http://www.ufoabduction.com/

Your Brain Hates You!

Skeptical Sidebar #9
How Far? How High? How Big? How Fast?

While we are on the subject of aliens let's take a moment to talk about UFOs, specifically some of the most common claims made about them and why, once you have a proper understanding of things, they will drive you doolally every time you hear them. Before we start let me clarify that when I'm talking about UFOs I mean Unidentified Flying Objects and not alien space craft from another dimension. An alien space craft would, by definition, be an Identified Flying Object, or IFO, as it has been identified as an alien space craft. A UFO is just that, an object in the sky that you cannot identify. Ok that little annoyance out of the way can you tell me what is wrong with this statement?

> "The first moving light we saw came from the east travelling at the height and speed similar to a Helicopter. I would estimate the distance to be 200 metres away...It would be

hard to estimate the size but my guess would be that of a small helicopter."[*]

So what's wrong with that? I'll give you a hint. Everything! Everything is wrong with that statement. Let's break it down shall we and show why you should never accept any claim about height, speed, distance or size when it comes to eyewitness accounts of UFO sightings. In order to do this let's conduct a rather odd thought experiment that I hope will make the problem clear to you.

I want you to imagine that you are standing on the second floor of a building and looking out of the window. Immediately outside there is a road, on the far side of which there is a wide field and then another similar road. From your position you can clearly see both roads and you know that the speed limit on both is 60 mph. Two cars approach, one on each road, both travelling exactly at the speed limit and precisely level with one another. How long, roughly, do you think it would take each car to cross your field of vision?

No, don't try to work out the maths, it is not important here. The simple answer is that the car closest to you will cross through your field of vision in about a second or two, while the car on the far side of the field will take maybe around ten seconds to make it all the way across. But how can this be if they are travelling at exactly the same speed? Well of course you know the answer to that, distance. One car is further away and therefore spends more time in your field of vision than the car that is close to you. As such if someone confirmed to you how fast each car was travelling I doubt you would have any trouble accepting that both

---

[*] Taken from an eyewitness statement of a UFO sighting that took place on 3rd June 2010 in Bromsgrove, Worcestershire in the UK. Is it just me or does it sound like he just saw a small helicopter?

Building your Skeptical Toolkit

vehicles are travelling the same speed, even though one is gone in a flash and the other takes a good few moments to disappear from sight. But let's take distance out of the equation shall we.

Imagine for the moment that you are Father Dougal from the hit UK comedy series Father Ted.* Despite Ted's best efforts to explain it you are still not understanding the difference between things that are small, and things that are far away. As such when you look out of the window at the two cars you come to the conclusion that the two cars are the same distance away from you and that one of them is just a lot smaller than the other one. With this bizarre notion in mind what conclusion would you come to about the speed of the two vehicles?

Well clearly the larger car crosses your field of vision much more quickly than the small car and so, based on the idea that they are both the same distance away, you would assume that the larger car was travelling faster than the smaller car. Of course you would be wrong.

Now what I hope this thought experiment makes clear is the idea that, when it comes to evaluating these things, speed, distance and size are very much interlinked. We'll ignore height for the moment and just focus on these three elements.

But very quickly let's address the subject of angles. As you read this book the chances are that it is probably no more than a foot or so away from your face. Holding it up in front of you at this distance you will notice that it takes up around a third of your field of vision. Lower the book and look across to the far side of the room at, let's say, the TV or some similarly sized object. Now your television is most likely a good deal larger than this book but you will notice that it only takes up around a tenth of your field of vision. Even though it is larger the TV covers a smaller angle in

---

* If you have never seen Father Ted then you really should. If you have seen it and didn't like it then, I'm sorry, but I am done with you.

417

your field of vision than the book simply due to the fact that it is further away from you. However, and this is the important part, because you know roughly how far away from you the TV is you are able to work out how large it is and so don't find yourself having Father Dougal's problem.

When it comes to UFOs this ability to make educated estimates about size and distance goes away because, by their nature, UFOs are found in the sky and the sky is a very different beasty indeed. When you look up into the sky at the clouds it is very difficult to work out how large they are because you simply have no context with which to compare them. The clouds could be massive and far away or smaller and lower to the ground. You simply can't tell and this is why when someone describes a UFO to you in terms of size and distance you should always be skeptical.

To work out the distance of an object you need to know how large the object is. To work out how large an object is you need to know how far away it is. These two elements are dependent upon one another. And as for speed, well we have already seen that this relies upon knowing how far the object is away from you as well. And with UFOs you don't know specific details about things like speed and size, after all that is kind of what the U means. Imagine that you are back at that second floor window and looking up into the sky. You see two similar objects that you can't identify travelling across the sky from right to left. Both objects appear to be the same size and travelling at the same speed, with the second object following close behind the first. However all is not as it seems.

The second object is actually ten times further away from you than the first one, ten times as large and travelling at ten times the speed. However because of this it appears to take up the same amount of space in your field of vision as the first object and also takes the same amount of time to move across that field of

Building your Skeptical Toolkit

vision as the first object. Simply put, a small object close to you and travelling slowly will appear exactly the same as a large object far away and travelling fast. And because these objects are "Unidentified" you don't know how large they are and so can't fill in the rest of the size/distance/speed equation.

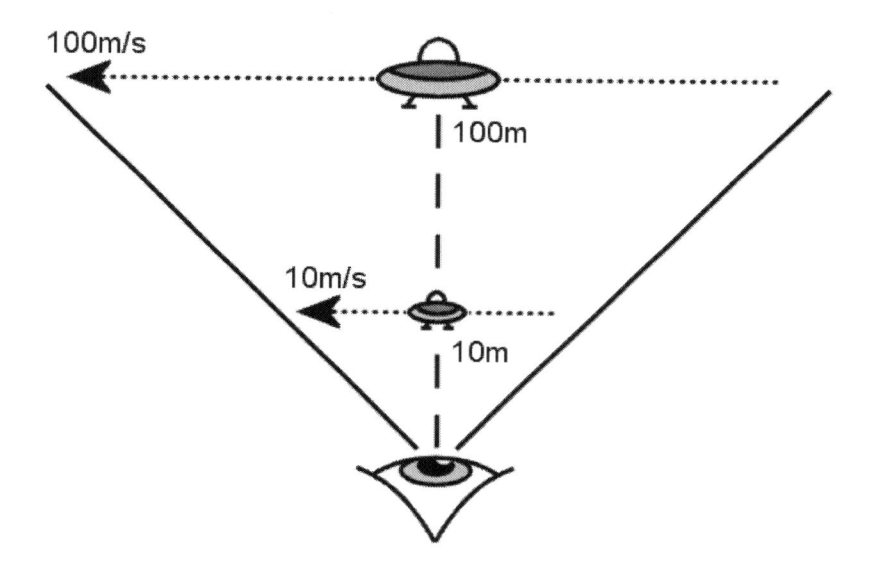

```
100m/s
◄·················     🛸     ·················
                     | 100m
                     |
                     |
        10m/s        |
        ◄········   🛸   ·········
                   | 10m
                     |
                    👁
```

But what about height, can you tell how far off the ground an object is? Let's make it easier for you. Let's say that just across the road from you is a tower than you know to be exactly ten meters tall. The two mysterious objects appear to be positioned slightly above the top of the tower. So does this mean you can take a guess at how high up they are? Well unsurprisingly the answer is no, as, once again, you need to take size and distance into the equation and we have already seen that we can't work these things out. On top of that angles really come into play here.

Your Brain Hates You!

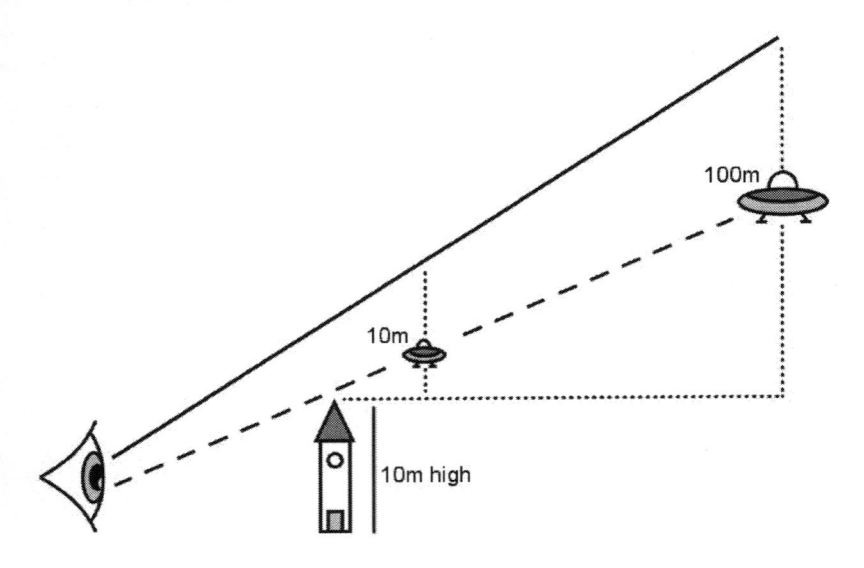

Because of your line of sight an object hovering just above and behind the top of the tower will appear to be at the same height above it as an object many times higher and many times further away. Don't believe me? Well then check it out for yourself. Hold one hand out in front of you with your thumb up. Now with the other hand hold your index finger out horizontally above your thumb and align it with the top edge of the wall across from you just where it meets the ceiling. If you are not in a room any high object will do. You will notice that your index finger is only a few inches above the top of your thumb and yet it appears to be at the same height as the top of the wall. Now unless you have a very odd room the top of the wall is many times higher above your thumb than your index finger, and yet they look the same.

The same thing goes for UFOs. Without knowing the size of the object or how far away it is from you it is simply impossible to work out how high up it is, even if you have an object of a known height between you and it. In fact even if the object passes directly in front of another object that is of a known size and

Building your Skeptical Toolkit

distance from you it doesn't help as, once again, a small object close to you and a large object far away will appear to be the same size. The best you could do would be to set a distance range somewhere between the tip of your nose and the known object and a compatible size range based on that. And yet somehow telling people that you saw something that was somewhere between the size of a bug and an ocean liner doesn't sound all that convincing, no matter how accurate it might be.

One final note on this subject. Just because someone is an airline pilot or in the army or even works for NASA doesn't mean that they have any special skill when it comes to judging the things we have just discussed. Yes a pilot may be able to accurately tell you the distance and speed of a 747, but that is because he knows the size of a 747 and thus can do the calculation needed to work out the rest. But with a UFO they do not have this information and without it, no matter how skilled they may be in other areas, any calculations they make are little more than guesswork.

Your Brain Hates You!

And in conclusion...
Beginning your skeptical journey

And so we reach the end of our time together, but hopefully just the beginning of your skeptical journey. Over the last few hundred pages I have presented you with numerous tools to help you on your way and now it's up to you to get out there and put them to good use. Before you go though I want to finish off by offering you a few words of advice on how to go about sharing your new found skeptical outlook on life and how you can be an effective light of reason to those around you. Now I have found that when it comes to the best way to express your skepticism to others there are many different opinions on how to go about it. As such it is likely some people will disagree with the advice I am about to give you, which is why I am starting with this important point:

Be Yourself

There is no 100% right, completely fool proof approach to being an effective skeptic. As such, first and foremost you should do what you think will work best for you and act in a way that you are comfortable acting. Not everyone wants to get into big debates or has the scientific knowledge to tackle a pseudoscientific claim head on, and there is nothing wrong with that. We all have different skills that we can bring to the table, find whatever it is you are good at and find a way to use it to spread the word of skepticism. That said however, here are a couple of things that I think you should all do.

Be the skeptic

One of the most effective ways I have found to share skepticism and critical thinking skills is to simply become known as the skeptic in the room. Now this doesn't mean forcing it down people's throats or boring them with it every waking hour of the day, and, as I said right at the beginning of this book, it also doesn't mean just disagreeing with people any time they mention they are seeing a chiropractor for their bad back or that their horoscope has put them in a good mood this morning. What it means is that when a topic comes up at work, or home or in your social life that warrants a bit of skepticism then, if appropriate to do so, be willing to offer up some critical thinking.

By employing the tools you have acquired from this book you should be able to present a reasoned and thoughtful opinion on many topics, even if they are completely new to you. So you have never heard of quantum meditative energy manipulation before[*] but you should have little trouble presenting a cogent argument as to why it is unlikely to be a reliable and effective form of medical intervention. Likewise you are probably not an expert in particle physics but that shouldn't stop you from correctly applying a healthy dose of skepticism when stories of faster than light neutrinos appear in the popular press[†]. Apply your skepticism often and you will quickly become known as the person to ask whenever a story pops up that seems a little odd or when someone wants a second opinion of whether they should

---

[*] And you won't have because I just made it up, though I did Google it just to make sure and came across an "interesting" website on Qigong meditation for energy manipulation. It seems you can't even make things up without someone having already thought it up and trying it for real - http://www.quantumjumping.com/blog/qigong-meditation/
[†] http://news.sciencemag.org/scienceinsider/2012/06/once-again-physicists-debunk.html

give ear candling a go. Of course to do this effectively does require a bit of work, which brings us to the next point.

Be informed

Skepticism, as with any skill, requires effort if you are going to get good at it. As such keep an eye out for chances to flex your skeptical muscle and always be on the lookout for opportunities to add a new tool to your kit. Over the course of this book I have only really scraped the surface of scientific and critical thinking skills, supplying you with just the main tools you are likely to need to address most issues that come your way. However, as with a skilled carpenter you will find there are times when the basic tools are not enough and a piece of specialised equipment is required. One of the best ways to acquire these specialised tools is to be on the lookout for stories that trigger your skeptical radar, but which you don't really know how to deal with. When they come up try to find out how other skeptics have handled them, paying careful attention to the methods they have used to deconstruct the claim. Often all that is required to do this is to add the word "skeptic" to a Google search on the topic and you will be gaining a new tool for your kit in no time.

Don't blind them with science

Once you have been applying your critical thinking skills for a while it is very easy to forget that not everyone has them, or at least doesn't use them as often as you do. As I've mentioned a number of times already skepticism doesn't come naturally, we have to work at it to overcome the natural way in which our brains work. As such when talking to someone about your reasons for being skeptical of some issue or explaining why you think a particular argument does not hold water it is always best not to overwhelm them with technical and scientific terminology. Of course this doesn't mean that you should talk down to them, but rather it means that instead of simply stating that something violates the 2nd law of thermodynamics or that it is utilising a non sequitur in its defence you should take the time to break down what these terms mean and explain your reasoning. For example you might point out that those who claim that the fact that moon rocks have been found in Antarctica is proof that we never went to the moon are employing a non sequitur argument because their conclusion, even if it were true, does not follow from their premise. Even if they disagree with you hopefully you will have been able to educate them a bit on logical fallacies rather than having them feel they have been lectured to.

Find common ground

Even when talking to someone who could be considered a true believer in some paranormal or pseudoscientific claim it is likely you can find something upon which you both agree. Use this area of common ground as the starting point of your conversation, as the foundation of your argument and to help you identify where your thinking diverges from theirs. Imagine you are talking to someone who believes in, to pick a fun example, Bigfoot. You might start by identifying a cryptid about which you are both skeptical, let's say the Loch Ness Monster. Get them to go through with you their reasons for not believe in Nessie as well as explaining your own reasons to doubt its existence and be sure to emphasis those points upon which you agree. Then use, where possible, these exact same arguments to explain your skepticism regarding the existence of Bigfoot. Alternatively you might ask them what evidence they would require in order to accept that the Loch Ness Monster was real and then see if the evidence they claim for the existence of Bigfoot measures up to their own standard. This is a far more effective approach than simply stating that they are wrong and is much more likely to give them something to think about.

Pick your battles

While we are on the subject of expressing your skepticism in a friendly and non-confrontational way it is worth mentioning that there are times when you should definitely keep your skepticism to yourself, or at least dial it down to a minimum. I mentioned this at the start of the book but it bears repeating. Sometimes there is simply no upside to expressing your skepticism on a given issue and everything to gain from, for the time being at least, keeping it to yourself. If the claim is harming no one, and may even be bringing them comfort, then it is often better to live and let live. A friend of mine who found herself faced with the prospect of having a loved one in harm's way started focusing again on her religious beliefs and the power of prayer to intervene on her behalf. This was most certainly not the time for me to act the hardened skeptic as doing so would most likely have pushed her away and made her less willing to listen to my thoughts on the topic at a later date. Sometimes you should just be a good friend and acknowledge that there are times when they require the support of something you yourself find difficult to accept. Of course I am not saying you can never broach the issue or that you can't politely and gently state that you do not share those beliefs, just that you don't need to go after every sacred cow that comes your way. Save your strength and ammunition for fights that really matter.

Your Brain Hates You!

Know your audience

All that said there are times when a more forthright approach is required. When it comes to expressing your skepticism openly you should always keep in mind what it is you are trying to accomplish and who you are aiming your arguments at. The person you are talking to may not always be the person you are trying to convince. If you find yourself talking to a diehard believer for instance then even your most powerful and well-reasoned arguments are unlikely to shake their convictions. However, that doesn't mean that putting your arguments out there is a waste of time if your aim is in fact to convince a third party to the validity of your reasoning and, by extension, the errors in your opponents thinking. Sometimes it is worth countering an argument simply to show that an alternative point of view exists. You may not convince anyone then and there but you never know what seeds you may have sown.

And of course this brings us to a point that often comes up in skeptical circles, the question of tone and the use of things like sarcasm and derision. Now I am a big supporter of the idea that if you can't say something nice then you shouldn't say anything at all. I also agreed wholeheartedly with skeptic, astronomer and former president of the James Randi Education Foundation, Phil Platt, when he succinctly summed up the issue by saying "don't be a dick". But then I also think that there are exceptions to every rule and once again it comes down to working out who your audience is. Yes, the old saying that you win more flies with honey is probably true, but then if your particular fly is so adamant in their belief that all the honey in the world is not going to change their mind then maybe another approach is in order. If a little derision and mockery will convince others that the proponent is in the wrong then I say go for it, though of course always back these

Building your Skeptical Toolkit

things up with evidence. While the friendly approach is, to my mind anyway, always the best place to start I don't think you should rule out other options. Use the style of debate that is appropriate for the situation and the audience you are trying to reach, whether it is the person you are talking to or some third party listening in.

Sow the seeds

I mentioned this in passing in the previous point but sometimes you don't need to win the argument or convince someone you are right then and there. Often it is more than enough just to sow some skeptical seeds and leave them to grow in their own time. Rather than confronting someone's beliefs head on simply explain that you are skeptical of, let's say, alien abductions due to the vast majority of such experiences being easily explained by well-established psychological conditions such as hypnagogia. Say that you would require more evidence to rule out the possibility that people are simply experiencing waking dreams and then maybe point out that anecdotal evidence itself is inherently unreliable and unscientific. Ask how they know that these issues have been addressed in the evidence they have presented. Then stand back and let the seeds grow.

Share the love

But of course skepticism isn't all about proving other people wrong and I hope I have managed to get across some of the beauty and wonder that comes along with scientific and critical thinking. If you are anything like me you will find real joy and passion in seeing this incredible universe of ours through the lenses of science and reason. If this is the case then share it with

429

others. Let them see your excitement and joy when you make a new discovery. Talk passionately about things that interest you, draw them in with your love of this miraculous reality around us. People may not share your level of interest in scientific thinking but when they see how you light up when talking about it they will want to know more. Psychologists have found that these kind of emotions are often infectious. For example if a complete stranger smiles at us we often instinctively smile back and doing so actually makes us feel happier. If you are enthusiastic about subjects that interest you then people will see them as subjects to be enthusiastic about.

Have fun

The other thing I hope I have got across in this book is that science, skepticism and critical thinking doesn't have to be boring. There are plenty of fun and enjoyable ways you can share skepticism with others, from magic tricks to pub games. So have fun with your skepticism, enjoy employing your critical thinking skills and delight in new scientific discoveries. And just because you are skeptical of the existence of dragons don't let that rob you of the fun that comes from the idea of dragons. Skepticism should enhance your life, make it fuller, more engaging and, importantly, more truthful. There is no reason it can't make it more fun as well.

So there you go, that is all I have to say on the topic, it is all up to you now. The tools you have been equipped with over the course of this book will stand you in good stead for the future and will let you see through the fog of modern life to the shining truths hiding beyond. But hey, don't take my word for it, be skeptical and check it out for yourself.

Bibliography and further reading

A religious fish. (n.d.). Retrieved July 5, 2012, from http://www.anomalies-unlimited.com/Religion/Allahfish.html.

Aaronovitch, D. (2009). Voodoo Histories – How conspiracy theory has shaped modern history. London: Random House. ISBN: 978-0099478966.

Acupuncture for back pain – 2009 update. (2009). Retrieved August 8th, 2012, from http://www.medicine.ox.ac.uk/bandolier/booth/painpag/Chronrev/Other/acuback.html.

Adams, M. (2010). Mumps outbreak spreads among people who got vaccinated against mumps. Retrieved June 28, 2012, from http://www.naturalnews.com/028142_mumps_vaccines.html.

Alicke, M.D., & Govorun, O. (2005). The better-than-average effect. In M.D. Alicke, D.A. Dunning, & J.I. Krueger (Eds.), The self in social judgment (pp. 85-106). New York: Psychology Press. ISBN: 978-1841694184.

All charges dropped in facilitated communication abuse case. (2008). Retrieved July 1, 2012, from http://www.baam.emich.edu/baammainpages/whtnew.htm#allchargesdropped.

Angel or Devil? Viewers see images in smoke. (2001). Retrieved July 5, 2012, from http://www.thedenverchannel.com/news/962839/detail.html.

Another acupuncture study misinterpreted. (2009). Retrieved August 8th, 2012, from http://scienceblogs.com/insolence/2009/05/13/another-acupuncture-study-misinterpreted/.

Aranza, J. (1987). Backward masking unmasked. Los Angeles: Vital Issues Press. ISBN: 978-0910311045.

Bibliography and further reading

Arnold, P. (2002). Easy Card Tricks. London: Octopus Publishing Group Ltd. ISBN: 978-0600607267.

Asch, S.E. (1946). Forming impressions of personality. Journal of Abnormal and Social Psychology, 41, 1230-1240.

Asch, S.E. (1951). Effects of group pressure upon the modification and distortion of judgements. In H. Guetzkow (Ed.). Groups, Leadership and Men (pp. 177-190). Pittsburgh: Carnegie Press.

Asch, S.E. (1956). Studies of independence and conformity: A minority of one against a unambiguous majority. Psychological Monographs: General and Applied, 70, 1-70. doi: 10.1037/h0093718.

Association TransCommunication: ATransC Theory. (n.d.). Retrieved July 6, 2012, from http://atransc.org/theory.htm.

Automatic writing. (n.d.). Retrieved July 1, 2012, from http://www.crystalinks.com/automatic_writing.html.

Ayton, P., & Fischer, I. (2004). The hot hand fallacy and the gambler's fallacy: Two faces of subjective randomness? Memory & Cognition, 32, 1369-1378.

Banaji, M.R., & Kihistrom, J.F. (1996). The ordinary nature of alien abduction memories. Psychological Inquiry, 7, 132-135. doi: 10.1207/s15327965pli0702_3.

Banks, J. (2001). Rorschach audio: Ghost voices and perceptual creativity. Leonardo Music Journal, 11, 77-83.

Baron, J. (2000). Thinking and deciding (3rd ed.). New York: Cambridge University Press. ISBN 0-521-65030-5

Barrow, J.D. (2008). 100 Essential Things You Didn't Know You Didn't Know. London: The Bodley Head. ISBN: 978-1847920034.

Baruss, I. (2001). Failure to replicate electronic voice phenomenon. Journal of Scientific Exploration, 15, 355-367.

Begg, I.M., Needham, D.R., & Bookbinder, M. (1993). Do backward messages unconsciously affect listeners? No. Canadian Journal of Experimental Psychology, 47, 1-14. doi: 10.1037/h0078772.

Bell, C.C., Shakoor, B., Thompson, B., Dew, D., Hughley, E., Mays, R., & Shorter-Gooden, K. (1984). Prevalence of isolated sleep paralysis in black subjects. Journal of the National Medical Association, 76, 501-508.

Berenbaum, M. (2005). Face time. American Entomologist, 51, 68-69.

Bernard, P. (2008). Toddler's Elmo doll makes death threats, family says. Retrieved July 5, 2012, from http://www2.tbo.com/news/offbeat/2008/feb/21/toddlers-elmo-doll-makes-death-threats-ar-150043/.

Berns, G.S., Chappelow, J., Zink, C.F., Pagnoni, G., Martin-Skurski, M.E., & Richards, J. (2005). Neurobiological correlates of social conformity and independence during mental rotation. Biological Psychiatry, 58, 245-253.

Bernstein, D.M, & Loftus, E.F. (2009). How to tell is a particular memory is true or false. Perspectives on Psychological Science, 4, 370-374.

Betz, H.D., Kulzer, R., König, H.L., Tritschler, J., & Wagner, H. (1996). Dowsing reviewed: The effect persists. Naturwissenschaften, 83, 272-275.

Blackmore, S. (1998). Abduction by aliens or sleep paralysis. Retrieved July 8, 2012, from http://www.csicop.org/si/show/abduction_by_aliens_or_sleep_paralysis.

Blackmore, S., & Cox, M. (2000). Alien abductions, sleep paralysis and the temporal lobe. European Journal of UFO and Abduction Studies, 1, 113-118.

Borgerson, K. (2009). Valuing evidence: Bias and the evidence hierarchy of evidence-based medicine. Perspectives in Biology and Medicine, 52, 218-233.

Borja, M.C., & Haigh, J. (2007). The birthday problem. Significance, 4, 124-127.

Boudry, M., Blancke, S., & Braeckman, J. (2010). How not to attack intelligent design creationism: Philosophical misconceptions about methodological naturalism. Foundations of Science, 15, 227-244. doi: 10.1007/s10699-010-9178-7.

Boynton, J. (2012). Facilitated communication – What harm it can do: Confessions of a former facilitator. Evidence-Based Communication Assessment and Intervention. iFirst Article, 1-11. doi: 10.1080/17489539.2012. 674680.

Brauer, M., Judd, C.M., & Gliner, M.D. (1995). The effects of repeated expressions on attitude polarization during group discussions. Journal of Personality and Social Psychology, 68, 1014-1029.

Bibliography and further reading

Braun, K.A., Ellis, R., & Loftus, E.F. (2002). Make my memory: How advertising can change our memories of the past. Psychology & Marketing, 19, 1-23.

Braun-LaTour, K.A., LaTour, M.S., Pickrell, J., & Loftus, E.F. (2004). How (and when) advertising can influence memory for consumer experience. Journal of Advertising, 33, 7–25.

Brehm, J.W., Cohen, A.R. (1962). Explorations in cognitive dissonance. Hoboken, NJ, US: John Wiley & Sons Inc. doi: 10.1037/11622-000

Brief report: Assessing allegations of sexual molestation made through facilitated communication. Journal of Autism and Developmental Disorders, 25, 319-326.

Brotchie, A. (n.d.). Automatic writing. Retrieved July 1, 2012, from http://www.duke.edu/web/lit132/automatic.html.

Brown, D. (2006). Tricks of the Mind. London: Channel 4 Books. ISBN: 978-1905026357.

Bryson, B. (2003). A Short History of Everything. London: Doubleday. ISBN: 978-0552997041.

Byrne, T., & Normand, M. (2000). The demon-haunted sentence: A skeptical analysis of reverse speech. Retrieved July 5, 2012, from http://www.csicop.org/si/show/demon-haunted_sentence_a_skeptical_analysis_of_reverse_speech1/.

Cariotto, M.J. (1988). Digital imagery analysis of unusual Martian surface features. Applied Optics, 27, 1926-1933. doi: 10.1364/AO.27.001926.

Carroll, R.T. (2003). The Skeptic's Dictionary: A Collection of Strange Beliefs, Amusing Deceptions, and Dangerous Delusions. Hoboken, New Jersey: John Wiley & Sons. ISBN: 978-0471272427.

Carroll, R.T. (2006). Backward satanic masking (backmasking). Retrieved July 5, 2012, from http://www.skepdic.com/backward.html.

Carroll, R.T. (2007). Oprah and the mother warriors against science. Retrieved June 17, 2012, from http://skepdic.com/autismthimerosal.html.

Carroll, R.T. (2010). Face on Mars. Retrieved July 3, 2012, from http://www.skepdic.com/faceonmars.html.

Carroll, R.T. (2010). Reverse speech. Retrieved July 5, 2012, from http://skepdic.com/reversespeech.html.

Carroll, R.T. (2011). Electronic voice phenomenon (EVP). Retrieved July 6, 2012, from http://www.skepdic.com/evp.html.

Carroll, R.T. (2011). Ideomotor effect. Retrieved June 28, 2012, from http://www.skepdic.com/ideomotor.html.

Carroll, R.T. (2011). Sleep paralysis. Retrieved July 7, 2012, from http://www.skepdic.com/sleepparalysis.html.

Carroll, R.T. (2012). Anti-vaccination movement (AVM). Retrieved June 17, 2012, from http://www.skepdic.com/antivaccination.html.

Carroll, R.T. (2012). Automatic writing (trace writing). Retrieved July 1, 2012, from http://www.skepdic.com/autowrite.html.

Carroll, R.T. (2012). Confirmation bias. Retrieved July 1, 2012, from http://www.skepdic.com/confirmbias.html.

Carroll, R.T. (2012). Facilitated communication (FC) ("supported typing"). Retrieved June 29, 2012, from http://www.skepdic.com/facilcom.html.

Carroll, R.T. (2012). Law of truly large numbers (coincidence). Retrieved June 27, 2012, from http://www.skepdic.com/lawofnumbers.html.

Carroll, R.T. (2012). Pareidolia. Retrieved July 3, 2012, from http://www.skepdic.com/pareidol.html.

Centers for disease control and prevention. National center for immunization and respiratory diseases. School vaccination coverage reports. (2008). Retrieved June 28, 2012, from http://www2.cdc.gov/nip/schoolsurv/nationalDetail8.asp?v=Mumps&y=2008.

Charpak, G., & Broch, H. (2004).Debunked!: ESP, telekinesis, and other pseudoscience. Baltimore, Maryland: The John Hopkins University Press. ISBN: 978-0801878671.

Cherkin, D.C., Sherman, K.J., Avins, A.L., Erro, J.H., Ichikawa, L., Barlow, W.E., ... Deyo, R.A. (2009). A randomized trial comparing acupuncture, simulated acupuncture and usual care for chronic low back pain. Archives of Internal Medicine, 169, 858-866.

Cherkin, D.C., Sherman, K.J., Hogeboom, C.J., Erro, J.H., Barlow, W.E., Deyo, R.A., & Avins, A.L. (2008). Efficacy of acupuncture for chronic low back pain: Protocol for a randomized controlled trial. Trials, 9, 1-11. doi:10.1186/1745-6215-9-10.

Bibliography and further reading

Cherry, K. (n.d.). Stages of sleep. Retrieved July 7, 2012, from http://psychology.about.com/od/statesofconsciousness/a/SleepStages.htm.

Cheyne, J. A. (2001). The ominous numinous. Sensed presence and 'other' hallucinations. Journal of Consciousness Studies, 8, 133-150.

Cheyne, J.A. (2005). Sleep paralysis episode frequency and number, types, and structure of associated hallucinations. Journal of Sleep Research, 14, 319-324. doi: 10.1111/j.1365-2869.2005.00477.x.

Cheyne, J.A., Rueffer, S.D., & Newby-Clark, I.R. (1999). Hypnagogic and hypnopompic hallucinations during sleep paralysis: Neurological and cultural construction of the night-mare. Consciousness & Cognition, 8, 319-337.

Chigwedere, P., Seage, G.R., Gruskin, S., Lee, T., & Essex, M. (2008). Estimating the lost benefits of antiretroviral drug use in South Africa. Journal of Acquired Immune Deficiency Syndromes, 49, 410-415. doi: 10.1097/QAI.0b013e31818a6cd5.

Choron, H., Choron, S. (2010). Look! It's Jesus! Seeing Jesus (and Mary) in food and other objects. Retrieved July 3, 2012, from http://www.beliefnet.com/Entertainment/Books/2010/02/Jesus-in-Food-and-Other-Objects.aspx.

Christianson, S., & Loftus, E.F. (1987). Memory for traumatic events. Applied Cognitive Psychology, 1, 225-239.

Clancy, S. A., McNally, R. J., Schacter, D. L., Lenzenweger, M. F., & Pitman, R. K. (2002). Memory distortion in people reporting abduction by aliens. Journal of Abnormal Psychology, 11, 455-461.

Clark, S.E., & Loftus, E.F. (1996). The construction of space alien abduction memories. Psychological Inquiry, 7, 140-143. doi: 10.1207/s15327965pli0702_5.

Cline, A. (n.d.). Burden of proof and fraud. People who make paranormal claims must prove those claims. Retrieved September 4th, 2012, from http://atheism.about.com/od/parapsychology/a/fraud.htm.

Cline, A. (n.d.). Flaws in reasoning and arguments: Confirmation bias. Retrieved July 1, 2012, from http://atheism.about.com/od/logicalflawsinreasoning/a/confirmation.htm.

Cline, A. (n.d.). Laws of logic. Retrieved July 20, 2012, from http://atheism.about.com/library/glossary/general/bldef_lawsoflogic.htm.

Cline, A. (n.d.). What are hypotheses, theories, facts in science? Retrieved July 13, 2012, from http://atheism.about.com/od/philosophyofscience/a/ScienceFacts.htm.

Cohen, L.J. (1981). Can human irrationality be experimentally demonstrated? The Behavioral and Brain Sciences, 4, 317-370.

Cold fusion: A case study for scientific behaviour. (n.d.). Retrieved July 20, 2012, from http://undsci.berkeley.edu/article/cold_fusion_01.

Confirmation bias. (n.d.). Retrieved July 1, 2012, from http://www.reference.com/browse/confirmation-bias.

Considine, J.D. (1995). Some lyrics are revised when listeners read between the lines changing their tune. Retrieved July 5, 2012, from http://articles.baltimoresun.com/1995-07-05/features/1995186124_1_lyrics-friends-are-jewish-album.

Cooper, C. (1989). Subliminal messages, heavy metal music and teen-age suicide. Retrieved July 5, 2012, from http://www.reversespeech.com/judas.htm.

Cooper, M.A. (n.d.). Medical aspects of lightning. Retrieved June 25, 2012, from http://www.lightningsafety.noaa.gov/medical.htm.

Cottrell, S. (2005). Palgrave study skills: Critical thinking skills. London: Palgrave Macmillan. ISBN: 978-1403996845.

Cox, B., & Forshaw, J. (2010). Why does E=mc2? Cambridge: Da Capo Press. ISBN: 978-0306819117.

Coyne, J.A. (2010). Methodological naturalism: does it exclude the supernatural? Retrieved July 20, 2012, from http://whyevolutionistrue.wordpress.com/2010/10/22/methodological-naturalism-does-it-exclude-the-supernatural/.

Coyne, J.A. (2010). Methodological naturalism: Does it exclude the supernatural? Retrieved September 4th, 2012, from http://whyevolutionistrue.wordpress.com/2010/10/22/methodological-naturalism-does-it-exclude-the-supernatural/.

Bibliography and further reading

Croson, R., & Sundali, J. (2005). The gambler's fallacy and the hot hand: Empirical data from casinos. The Journal of Risk and Uncertainty, 30, 195-209.

Cucherat, M., Haugh, M.C., Gooch, M., & Boissel, J.P. (2000). Evidence of clinical efficacy of homeopathy: A meta-analysis of clinical trials. European Journal of Clinical Pharmacology, 56, 27-33. doi: 10.1007/s002280050716.

Dagnall, N., Drinkwater, K., & Parker, A. (2011). Alien visitation, extra-terrestrial life, and paranormal beliefs. Journal of Scientific Exploration, 25, 699-720.

Dagnall, N., Munley, G., Parker, A., & Drinkwater, K. (2010). The relationship between belief in extra-terrestrial life, UFOs-related beliefs and paranormal belief. Society for Psychical Research, 74, 1–14.

Dahlitz, M., & Parkes, J.D. (1993). Sleep paralysis. Lancet, 341, 406-407.

Daly, J., Willis, K., Small, R., Green, J., Welch, N., Kealy, M., & Hughes, E. (2007). A hierarchy of evidence for assessing qualitative health research. Journal of Clinical Epidemiology, 60, 43-49.

Daniels, M. (n.d.). Electronic voice phenomena. Retrieved July 6, 2012, from http://www.psychicscience.org/evp.aspx.

Davies, G. (2009). Estimating the speed of vehicles: the influence of stereotypes. Psychology, Crime & Law, 15, 293-312 doi: 10.1080/10683160802203971.

DeMarche, E., Falco, M., & Bliman, N. (2010). More than 1,000 get mumps in New York, New Jersy since August. Retrieved June 28, 2012, from http://edition.cnn.com/2010/HEALTH/02/08/mumps.outbreak.northeast/index.html?hpt=T2.

Depue, B.E. (2010). False Memory Syndrome. Corsini Encyclopedia of Psychology. 1–2. doi: 10.1002/9780470479216.corpsy0346.

Dering, B., Martin, C.D., Moro, S., Pegna, A.J., & Thierry, G. (2011). Face-sensitive processes one hundred milliseconds after picture onset. Frontiers in Human Neurosceicne, 5, 1-14.

Devil face in smoke of 9/11 at the WTC. (n.d.). Retrieved July 5, 2012, from http://www.christianmedia.us/devil-face.html.

Dewey, R.A. (2007). REM sleep in cats. Retrieved July 7, 2012, from http://www.psywww.com/intropsych/ch03_states/rem_sleep_in_cats.html.

Diaz, D., (2009). Foul-mouthed Teletubby doll? Retrieved July 5, 2012, from http://www.cbsnews.com/2100-500609_162-11282.html.

Dickerson, B. (2008). Part two: Sex abuse case against Oakland couple was legal horror show. Retrieved July 1, 2012, from http://www.freep.com/apps/pbcs.dll/article?AID=/20080317/COL04/803170336.

Dolgin, E. (2009). Publication bias continues despite clinical-trial registration. Retrieved August 8th, 2012, from http://www.nature.com/news/2009/090911/full/news.2009.902.html.

Dorko, B.L. (n.d.). Without volition: The presence and purpose of ideomotor movements. Retrieved June 28, 2012, from http://www.barrettdorko.com/articles/ideomotor.htm.

Dunning, B. (2008). Betty and Barney Hill: The original UFO abduction. Retrieved July 9, 2012, from http://skeptoid.com/episodes/4124.

Dunning, B. (2008). Skeptoid 2: More Critical Analysis of Pop Phenomena. Skeptoid Media Inc. ISBN: 978-1-440422850.

Dunning, B. (2008). Skeptoid: Critical Analysis of Pop Phenomena. CreateSpace. ISBN: 978-1434821669.

Dunning, B. (2008). The face on Mars revealed. Retrieved July 3, 2012, from http://skeptoid.com/episodes/4097.

Dunning, B. (2008). When people talk backwards. Retrieved July 3, 2012, from http://skeptoid.com/episodes/4105.

Dworschak, M. (2010). Communication with those trapped within their brains. Retrieved July 1, 2012, from http://www.spiegel.de/international/world/neurological-rescue-mission-communicating-with-those-trapped-within-their-brains-a-677537.html.

Eller, S.G. (n.d.). Automatic writing. Retrieved July 1, 2012, from http://www.healingfromwithin.com/Articles/Automatic_Writing/automatic_writing.html.

Ellis, K.M. (1995). The Monty Hall problem. Retrieved June 28, 2012, from http://montyhallproblem.com/.

Bibliography and further reading

Emery, D. (2008). Urban legends: Talking doll allegedly says 'Islam is the light'. Retrieved July 5, 2012, from http://urbanlegends.about.com/b/2008/10/10/talking-doll-allegedly-says-islam-is-the-light.htm.

Enright, J.T. (1995). Water dowsing: the Scheunen experiments. Naturwissenschaften, 82, 360-369.

Enright, J.T. (1999). Testing dowsing: The failure of the Munich experiments. Retrieved June 28, 2012, from http://www.csicop.org/si/show/testing_dowsing_the_failure_of_the_munich_experiments/.

Eppig, E. (2010). Debunking ghosts: Hypnagogia and hypnopompia. Retrieved July 7, 2012, from http://suite101.com/article/debunking-ghosts-hypnagogia-and-hypnopompia-a272915.

Ernst, E. (2011). Acupuncture use by midwives. Retrieved September 3rd, 2012, from http://www.pulsetoday.co.uk/comment-blogs/-/blogs/12377499/acunpuncture-use-by-midwives.

Evans, D. (2003). Hierarchy of evidence: A framework for ranking evidence evaluating healthcare interventions. Journal of Clinical Nursing, 12, 77-84.

Evidence Based Nursing Practice: The Hierarchy of Evidence. (n.d.). Retrieved August 6th, 2012, from http://www.ebnp.co.uk/The%20Hierarchy%20of%20Evidence.htm.

EVP – Electronic voice phenomenon. (n.d.). Retrieved July 6, 2012, from http://www.trueghosttales.com/evp.php.

EVP equipment & recording methods. (n.d.). Retrieved July 6, 2012, from http://www.evpuk.com/evp_equipment.html.

Facilitated communication: Sifting the psychological wheat from the chaff. (2003). Retrieved June 29. 2012, from http://www.apa.org/research/action/facilitated.aspx.

Favere-Marchesi, M. (2006). "Order effects" revisited: The importance of chronology. Retrieved July 1, 2012, from http://aaahq.org/audit/midyear/06midyear/papers/favere_marchesi.pdf.

Festinger, L. (1957). A theory of cognitive dissonance. Stanford, CA: Stanford University Press.

Festinger, L., & Carlsmith, J. M. (1959). Cognitive consequences of forced compliance. Journal of Abnormal and Social Psychology, 58, 203-211.

Festinger, L., Riechen, H.W, & Schachter, S. (1956). When prophecy fails: A social and psychological study of a modern group that predicted the destruction of the world. New York: Harper-Torchbooks. ISBN: 978-0061311321.

Findlay, E. (2011). Other F-Word victims: Teletubbies, Pogues and UK "Faggots" ad. Retrieve July 5, 2012, from http://suite101.com/article/other-f-word-victims-teletubbies-pogues-and-uk-faggots-ad-a336389.

Finley, W. (2010). Sleep paralysis: From incubi to aliens. Retrieved July 7, 2012, from http://wilfinley.com/2010/01/25/sleep-paralysis-from-incubi-to-aliens/.

Firestone, M. (1985). The "Old Hag": Sleep paralysis in Newfoundland. Journal of Psychoanalytic Anthropology, 8, 47-66.

Folstein, S.E., & Piven, J. (1991). Etiology of autism: Genetic influences. Pediatrics, 87, 767-773.

Forgas, J.P., Laham, S.M., & Vargas, P.T. (2005). Mood effects on eyewitness memory: Affective influences on susceptibility to misinformation. Journal of Experimental Social Psychology, 41, 574-588. doi: 10.1016/j.jesp.2004.11.005.

Foster, J.L., Huthwaite, T., Yesberg, J.A., Garry, M., & Loftus, E.F. (2012). Repetition, not number of sources, increases both susceptibility to misinformation and confidence in the accuracy of eyewitnesses. Acta Psychologica, 139, 320-326. doi: 10.1016/j.actpsy.2011.12.004.

French, C. (2009). Close encounters of the faked kind. Retrieved July 7, 2012, from http://www.guardian.co.uk/science/2009/nov/09/the-fourth-kind-sleep-paralysis.

French, C. (2009). Families are still living the nightmare of false memories of sexual abuse. Retrieved June, 23, 2012, from http://www.guardian.co.uk/science/2009/apr/07/sexual-abuse-false-memory-syndrome.

Bibliography and further reading

French, C. (2009). The waking nightmare of sleep paralysis. Retrieved July 7, 2012, from http://www.guardian.co.uk/science/2009/oct/02/sleep-paralysis.

French, C., & Wilson, K. (2006). Incredible Memories — How Accurate are Reports of Anomalous Events? European Journal of Parapsychology, 21.2, 166–181.

French, C.C., & Richards, A. (1993). Clock this!: An everyday example of a schema-driven error in memory. British Journal of Psychology, 84, 249–253. doi: 10.1111/j.2044-8295.1993.tb02477.x.

French, C.C., Santomauro, J., Hamilton, V., & Fox, R. (2009). Psychological aspects of the alien contact experience. In M.D. Smith (Ed). Anomalous experiences: Essays from parapsychological and psychological perspectives. (pp 136-153). Jefferson, North Carolina: McFarland & Company. ISBN: 978-0786443987.

Friedman, M.H., & Weisberg, J. (1981). Applied kinesiology—double-blind pilot study. Journal of Prosthetic Dentistry, 45, 321–323.

Frye, T. (2008). Pareidolia. Retrieved July 3, 2012, from http://www.weird-encyclopedia.com/pareidolia.php.

Geraerts, E., Bernstein, D.M., Merckelbach, H., Linders, C., Raymaekers, L., & Loftus E.F. (2008). Lasting false beliefs and their behavioural consequences. Psychological Science, 19, 749-753.

Gianotti, L.R.R., Mohr, C., Pizzagalli, D., Lehmann, D., & Brugger, P. (2001). Associative processing and paranormal belief. Psychiatry and Clinical Neuroscience, 55, 595-603.

Godlee, F. (2011). The fraud behind the MMR scare. British Medical Journal, 342, d22. doi: 10.1136/bmj.d22.

Godlee, F., Smith, J., & Marcovitch, H. (2011). Wakefield's article linking MMR vaccine and autism was fraudulent. British Medical Journal, 342, c7453. doi: 10.1136/bmj.c7452.

Goldacre, B. (2007). A menace to science. Retrieved August 31st, 2012, from http://www.guardian.co.uk/media/2007/feb/12/advertising.food.

Goldacre, B. (2008). All bow before the might of the placebo effect, it is the coolest strangest thing in medicine. Retrieved August 8th, 2012, from http://www.badscience.net/2008/03/all-bow-before-the-might-of-the-placebo-effect-it-is-the-coolest-strangest-thing-in-medicine/.

Goldacre, B. (2008). Bad Science. London: Fourth Estate. ISBN: 978-0007240197.

Goode, E., & Ben-Yahuda, N. (1994). Moral panics: Culture, politics, and social construction. Annual Review of Sociology, 20, 149-171.

Goren, C.C., Sarty, M., & Wu, P.Y.K. (1975). Visual following and pattern discrimination of face-like stimuli by newborn infants. Pediatrics, 56, 544-549.

Gorski, D. (2009). Welcome back, my friends, to the show that never ends: The Jenny and Jim antivaccine propaganda tour has begun. Retrieved June 17, 2012, from http://www.sciencebasedmedicine.org/index.php/the-jenny-and-jim-antivaccine-propaganda-tour-has-begun/.

Gould, S.J. (1981). Evolution as fact and theory. Discover, 2, 34-37.

Government statement on 'bomb detectors' export ban. (2010). Retrieved June 17, 2012, from http://news.bbc.co.uk/1/hi/programmes/newsnight/8475875.stm.

Gow, K. M., Hutchinson, L., & Chant, D. (2009). Correlations between fantasy proneness, dissociation, personality factors and paranormal beliefs in experiencers of paranormal and anomalous phenomena. Australian Journal of Clinical Experimental Hypnosis, 37, 169-191.

Greenhalgh, T. (1997). How to read a paper: Getting your bearings (deciding what the paper is about). British Medical Journal, 515, 243-246.

Grobman, K.H. (n.d.). Teaching about confirmation bias. Retrieved July 1, 2012, from http://www.devpsy.org/teaching/method/confirmation_bias.html.

Grothe, D.J. (2012). "Psychic" Sally Morgan sues critics for £150,000 after refusing $1 million to prove her powers. Retrieved June 23, 2012, from http://www.huffingtonpost.com/dj-grothe/psychic-sally-morgan-sues_b_1244151.html.

Guide to clinical preventive services: Report of the U.S. Preventive Services Task Force. (1989). Darby, Pennsylvania: Diane Publishing Co.

Bibliography and further reading

Hadjikhani, N., Kveraga, K., Naik, P., & Ahlfors, S.P. (2009). Early (N170) activation of face-specific cortex by face-like objects. Neuroreport, 20, 403-407. doi: 10.1097/WNR.0b013e328325a8e1.

Hansen, G.P. (1982). Dowsing: A review of experimental research. The Journal of the Society for Psychical Research, 51, 343-367.

Harold Camping admits he was wrong about end of world prediction. (2012). Retrieved June 23, 2012, from http://www.huffingtonpost.com/2012/03/09/harold-camping-admits-hes-wrong_n_1335232.html.

Harold Camping prophecy leads to mothers attempted murder of children. (2011). Retrieved June 23, 2012, from http://www.liveleak.com/view?i=be6_1306075236.

Harold Camping says end did come May 21, spiritually; Predicts new date: October 21. (2011). Retrieved June 23, 2012, from http://au.ibtimes.com/articles/150707/20110524/harold-camping-says-end-did-come-may-21-spiritually-predicts-new-date-october-21.htm.

Harold Camping: Doomsday prophet wrong again. (2011). Retrieved June 23, 2012, from http://abcnews.go.com/blogs/business/2011/10/harold-camping-doomsday-prophet-wrong-again/.

Hawley, C., & Jones, M. (2010). Export ban for useless 'bomb detector'. Retrieved June 28, 2012, from http://news.bbc.co.uk/1/hi/8471187.stm.

Herbert, I. (2009). The psychology and power of false confessions. Retrieved July 1, 2012, from http://www.psychologicalscience.org/observer/getArticle.cfm?id=2590.

Herbert, W. (2012). Law and disorder: The psychology of false confession. Retrieved July 1, 2012, from http://www.psychologicalscience.org/index.php/news/were-only-human/in-the-criminal-justice-system.html.

Hilton, S., Petticrew, M., & Hunt, K. (2007). Parents' champions vs. vested interests: Who do parents believe about MMR? A qualitative study. BMC Public Health, 7, 42. doi: 10.1186/1471-2458-7-42.

Holden, K.J., & French, C.C. (2002). Alien abduction experiences: Some clues from neuropsychology and neuropsychiatry. Cognitive Neuropsychiatry, 7, 163-178. doi: 10.1080/13546800244000058.

Holden, S. (1983). Serious issues underlie a new album from Styx. Retrieved July 5, 2012, from http://www.nytimes.com/1983/03/27/arts/serious-issues-underlie-a-new-album-from-styx.html.

Holguin, J. (2006). Backmasking unmasked! Music site's in heavy rotation. Retrieved July 5, 2012, from http://seattletimes.nwsource.com/html/entertainment/2002828774_backmasking27.html.

Holliday, R.E., Douglas, K.M., & Hayes, B.K. (1999). Children's eyewitness suggestibility: Memory trace strength revisited. Cognitive Development, 14, 443-462.

Hood, B. (2009). SuperSense. London: HarperOne. ISBN: 978-1849010306.

Hornig, M., Briese, T., Buie, T., Bauman, M.L., Lauwers, G., et al. (2008) Lack of association between measles virus vaccine and autism with enteropathy: A case-control study. PLoS ONE, 3, e3140. doi: 10.1371/journal.pone.0003140.

How 'mommy instinct' outdid science (2011). Retrieved June 17, 2012, from http://www.smh.com.au/world/science/how-mommy-instinct-outdid-science-20110204-1agui.html.

How to get rid of an incubus demon? (2007). Retrieved July 8, 2012, from http://answers.yahoo.com/question/index?qid=20071217092429AAONPWA.

How to make your own EVP recordings. (n.d.). Retrieved July 6, 2012, from http://www.trueghosttales.com/how-to-record-evp.php.

Hviid, A., Stellfeld, M., Wohlfahrt, J., & Melbye, M. (2003). Association between thimerosal-containing vaccine and autism. The Journal of the American Medical Association, 290, 1763-1766. doi: 10.1001/jama.290.13.1763.

Hyman, R. (1996). Dowsing. In G. Stein (Ed.), The Encyclopedia of the Paranormal (pp. 222-233). Buffalo, New York: Prometheus Books.

Bibliography and further reading

Hyman, R. (2003). How people are fooled by ideomotor action. Retrieved June 28, 2012, from http://www.quackwatch.org/01QuackeryRelatedTopics/ideomotor.html.

Ibuprofen may raise risk of heart attack. (2005). Retrieved June 28, 2012, from http://www.medicalnewstoday.com/releases/25953.php.

Ichikawa, H., Kanazawa, S., Yamaguchi, M.K. (2011). Finding a face in a face-like object. Perception, 40, 500-502.

Isaak, M. (2004). The second law of thermodynamics says that everything tends toward disorder, making evolutionary development impossible. Retrieved August 27th, 2012, from http://www.talkorigins.org/indexcc/CF/CF001.html.

Isenberg, D.J. (1986). Group polarization: A critical review and meta-analysis. Journal of Personality and Social Psychology, 50, 1141-1151.

Jacobs, D.M. (1999). The Threat. New York: Fireside. ISBN: 978-0684848136.

Jacobson, J. W., Mulick, J. A., & Schwartz, A. A. (1995). A history of facilitated communication: Science, pseudoscience, and antiscience. (Science Working Group on facilitated communication). American Psychologist, 50, 750-765.

Jacobson, M. (2011). Satan's face. Retrieved July 5, 2012, from http://nymag.com/news/9-11/10th-anniversary/satans-face/.

Jarrett, C. (2012). Memory myths. Retrieved June 23, 2012, from http://www.guardian.co.uk/lifeandstyle/2012/jan/14/truth-about-memories-jarrett

Johnson, M.H., Dziurawiec, S., Ellis, H.D., & Morton, J. (1991). Newborns' preferential tracking of face-like stimuli and its subsequent decline. Cognition, 40, 1-19.

Johnson-Laird, P.N., & Wason, P.C. (1977). Thinking: Reading in cognitive science. Cambridge: Cambridge University Press. ISBN: 978-0521217569.

Jones, M., & Sugden, R. (2001). Positive confirmation bias in the acquisition of knowledge. Theory and Decision, 50, 59-99. doi: 10.1023/A:1005296023424.

Judas Priest suicide trial article. (2006). Retrieved July 5, 2012 from http://mahou.wordpress.com/2006/12/01/judas-priest-suicide-trial-article/.

Kaasa, S.O., Morris, E.K., & Loftus, E.F. (2011). Remembering why: Can people consistently recall reasons for their behaviour? Applied Cognitive Psychology, 25, 35-42. doi: 10.1002/acp.1639.

Kaku, M. (2005). Parallel Worlds. London: Penguin Books. ISBN: 978-0141014630.

Kane, W. (2011). Harold Camping 'flabbergasted'; rapture a no-show. Retrieved June 23, 2012, from http://www.sfgate.com/bayarea/article/Harold-Camping-flabbergasted-rapture-a-no-show-2370758.php.

Kaptchuk, T.J. (2003). Effect of interpretive bias on research evidence. British Medical Journal, 326, 1453-1455. doi: 10.1136/bmj.326.7404.1453.

Kassin, S.M., Goldstein, C.C., & Savitsky, K. (2003). Behavioral confirmation in the interrogation room: On the dangers of presuming guilt. Law and Human Behavior, 27, 187-203.

Kassin, S.M., Meissner, C.A., & Norwick, R.J. (2005). "I'd know a false confession if I saw one": A comparative study of college students and police investigators. Law and Human Behavior, 29, 211-227. doi: 10.1007/s10979-005-2416-9.

Kenny, J.J., Clemens, R., & Forsythe, K.D. (1988). Applied Kinesiology Unreliable for Assessing Nutrient Status. Journal of the American Dietetic Association, 88, 698-704.

Kimura, D., & Folb, S. (1968). Neural processing of backwards-speech sounds. Science, 161, 395-396.

Kingdom, F.A.A., Yoonessi, A., & Gheorghiu, E. (2007). The leaning tower illusion: A new illusion of perspective. Perception, 36, 475-477. doi: 10.1068/p5722a.

Krauss, L.M. (1996). The Physics of Star Trek. London: Harper Collins Publishers. ISBN: 978-0006550426.

Kreiner, D.S., Altis, N.A., & Voss, C.W. (2003). A test of the effect of reverse speech on priming. The Journal of Psychology: Interdisciplinary and Applied, 137, 224-232. doi: 10.1080/00223980309600610.

Bibliography and further reading

Kreiner, D.S., Price, R.Z., Gross, A.M., & Appleby, K.L. (2004). False recall does no increase when words are presented in a gender-congruent voice. Journal of Articles in Support of the Null Hypothesis, 3, 1-18.

Kreinovich, V., & Iourinski, D. (2002). Was there Satan's face in the World Trade Center fire? A geometric analysis. Geombinatorics, 12, 69-75.

Kuhn, D., & Lao, J. (1996). Effects of evidence on attitudes: Is polarization the norm? Psychological Science, 7, 115-120. doi: 10.1111/j.1467-9280.1996.tb00340.x.

Kumar, M. (2008). Quantum: Einstein, Bohr and the Great Debate about the Nature of Reality. London: Icon Books. ISBN: 978-19848310353.

Kunda, Z. (1990). The case form motivated reasoning. Psychological Bulletin, 108, 480-498. doi: 10.1037/0033-2909.108.3.480.

Kurus, M. (n.d.). Auras, chakras and energy fields: Cleansing and activating your energy systems. Retrieved August 27th, 2012, from http://www.mkprojects.com/fa_CleansingAndEnergizing.htm.

Lambert, F.L. (2002). Disorder – A cracked crutch for supporting entropy discussions. The Journal of Chemical Education, 79, 187. doi: 10.1021/ed079p187.

Lanford, A., & Lanford, J. (n.d.). About the Nigerian scam (Nigerian advanced fee scam): Internet ScamBusters #11. Retrieved June 17, 2012, from http://www.scambusters.org/NigerianFee.html.

Last year's Apocalyptic predictions were wrong, preacher admits. (2012). Retrieved June 23, 2012, from http://www.nj.com/news/index.ssf/2012/03/end_of_world_predictions_were.html.

Law of identify. (n.d.). Retrieved July 20, 2012, from http://objectivism101.com/Lectures/Lecture19.shtml.

Leach, B. (2009). Boots: 'we sell homeopathic remedies because they sell, not because they work'. Retrieved June 17, 2012, from http://www.telegraph.co.uk/finance/newsbysector/retailandconsumer/6658864/Boots-we-sell-homeopathic-remedies-because-they-sell-not-because-they-work.html.

Levine, R., & Evers, C. (n.d.). The slow death of spontaneous generation (1668-1859). Retrieved July 13, 2012, from http://www.accessexcellence.org/RC/AB/BC/Spontaneous_Generation.php.

Lewis. P. (2006). A fish called Allah. Retrieved July 5, 2012, from http://www.guardian.co.uk/uk/2006/feb/02/paullewis.

Lilienfeld, S.O., Lynn, S.J., Ruscio, J., & Beyerstein, B.L. (2009). 50 Great Myths of Popular Psychology: Shattering Widespread Misconceptions about Human Behavior. Chichester, West Sussex: Wiley-Blackwell. ISBN: 978-1405131124.

Lindermann, D.L. (1998). Reverse speech analysis: Interview with David John Oates. Retrieved July 5, 2012, from http://www.reversespeech.com/interview1wdjo.htm.

List of backmasked messages. (2012). Retrieved July 5, 2012, from http://en.wikipedia.org/wiki/List_of_backmasked_messages.

Lloyd, J., & Mitchinson, J. (2006). The Book of General Ignorance. London: Faber and Faber Limited. ISBN: 978-0571233687.

Loftus, E. (1997). Creating false memories. Scientific American, 277, 70-75.

Loftus, E. (2003). Our changeable memories: legal and practical implications. Nature Reviews, 4, 231-234. doi: 10.1038/nrn1054.

Loftus, E.F. (1979). The malleability of human memory: Information introduced after we view an incident can transform memory. American Scientist, 67, 312-320.

Loftus, E.F. (2011). Intelligence gathering post-9/11. American Psychologist, 66, 532-541. doi: 10.1037/a0024614.

Loftus, E.F., & Hoffman, H.G. (1989). Misinformation and memory: The creation of new memories. Journal of Experimental Psychology, 118, 100-104.

Loftus, E.F., Miller, D.G., & Burns, H.J. (1978). Semantic integration of verbal information into a visual memory. Journal of Experimental Psychology, 4, 19-31.

Loftus, G.R. (2010). What can a perception-memory expert tell a jury? Psychonomic Bulletin & Review, 17, 143-148. doi:10.3758/PBR.17.2.143.

Loftus. E.F. (1992). When a lie becomes memory's truth: Memory distortion after exposure to misinformation. Current Directions in Psychological Science, 1, 121-123.

Bibliography and further reading

Loftus. E.F. (2005). Planting misinformation in the human mind: A 30-year investigation of the malleability of memory. Learning & Memory, 12, 361-366. doi: 10.1101/lm.94705.

Lord, C.G., Ross, L., & Lepper, M.R. (1979). Biased assimilation and attitude polarization: The effects of prior theories on subsequently considered evidence. Journal of Personality and Social Psychology, 37, 2098-2109. doi: 10.1037/0022-3514.37.11.2098.

Lüdtke, R., Kunz, B., Seeber, N., & Ring, J. (2001). Test-retest-reliability and validity of the Kinesiology muscle test. Complementary Therapies in Medicine, 9, 141-145.

Lungu, C. and Hallett, M. (2011) Other Jerks and Startles. In A. Albanese & J. Jankovic (Eds.), Hyperkinetic Movement Disorders: Differential Diagnosis and Treatment. (pp. 236-256). Oxford, UK: Wiley-Blackwell. doi: 10.1002/9781444346183.ch16.

Lynn, S.J., & Kirsh, I.I. (1996). Alleged alien abductions: false memories, hypnosis and fantasy proneness. Psychological Inquiry, 7, 151-155. doi: 10.1207/s15327965pli0702_8.

MAAR Abduction Data. (2009). Retrieved July 9, 2012, from http://www.maar.us/abduction_data.html.

Malin Space Science Systems, Inc. (1995). The "Face on Mars". Retrieved July 3, 2012, from http://www.msss.com/education/facepage/face.html.

Maquet, P., Ruby, P., Maudoux, A., Albouy, G., Sterpenich, V., Dang-Vu, T., ... Laureys, S. (2005). Human cognition during REM sleep and the activity profile within frontal and parietal corticies: A reappraisal of functioning neuroimaging data. Progress in Brain Research, 150, 219-227. doi: 10.1016/S0079-6123(05)50016-5.

Marmite Messiah and a collection of pareidolia. (2009). Retrieved July 3, 2012, from http://icbseverywhere.com/blog/first-years-archives/marmite-messiah-collection-pareidolia/.

Martin, B. (1998). Coincidences: Remarkable or random? Retrieved June 27. 2012, from http://www.csicop.org/si/show/coincidences_remarkable_or_random/.

Mathieu, S., Boutron, I., Moher, D., Altman, D.G., & Ravaud, P. (2009). Comparison of registered and published primary outcomes in randomized controlled trials. The Journal of the American Medical Association, 302, 977-984.

McCarty, D.E., & Chesson, A.L. (2009). A case of sleep paralysis with hypnopompic hallucinations. Journal of Clinical Sleep Medicine, 5, 83-84.

McNally, R. J., & Clancy, S. A. (2005). Sleep paralysis, sexual abuse, and space alien abduction. Transcultural Psychiatry, 42, 113-122. doi: 10.1177/1363461505050715.

McNally, R.J., Lasko, N.B., Clancy, S.A., Macklin, M.L., Pitman, R.K., & Orr, S.P. (2004). Psychophysiological responding during script-driven imagery in people reporting abduction by space aliens. Psychological Science, 15, 493-497.

McNerney, S. (2011). Psychology's treacherous trio: Confirmation bias, cognitive dissonance, and motivated reasoning. Retrieved July 1, 2012, from http://whywereason.com/2011/09/07/psychologys-treacherous-trio-confirmation-bias-cognitive-dissonance-and-motivated-reasoning/.

Merlin, T., Weston, A., & Tooher, R. (2009). Extending an evidence hierarchy to include topics other than treatment: Revising the Australian 'levels of evidence'. BMC Medical Research Methodology, 9:34, 1-8. doi:10.1186/1471-2288-9-34.

Milgrom, L.R. (2005). Are randomized controlled trials (RCTs) redundant for testing the efficacy of homeopathy? A critique of RCT methodology based on entanglement theory. The Journal of Alternative and Complementary Medicine, 11, 831-838.

Miller, G.A. (1956). The magical number seven, plus or minus two: Some limits on our capacity for processing information. The Psychological Review, 63, 81-97.

Miller, M.B., Valsangkar-Smyth, M., Newman, S., Dumont, H., & Wolford, G. (2005). Brain activations associated with probability matching. Neuropsychologia, 43, 1598-1608. doi: 10.1016/j.neuropsychologia.2005.01.021.

Mills, M. (2007). Hidden and Satanic messages in rock music. Retrieved July 5, 2012, from http://blog.wfmu.org/freeform/2007/01/365_days_1_the_.html.

Bibliography and further reading

Miracles of almighty Allah. (n.d.). Retrieved July 5, 2012, from http://www.ya-hussain.com/int_col1/int_coll_net/miracles/miracles_of_almighty_allah.htm.

Moerman, D.E., & Harrington, A. (2005). Making space for the placebo effect in pain medicine. Seminars in Pain Medicine, 3, 2-6.

Mooney, C. (2005). Waking up to sleep paralysis. Retrieved July 7, 2012, from http://www.csicop.org/specialarticles/show/waking_up_to_sleep_paralysis/.

Moore, M. (2008). Talking Fisher-Price doll accused of promoting Islam. Retrieved July 5, 2012, from http://www.telegraph.co.uk/news/3191347/Talking-Fisher-Price-doll-accused-of-promoting-Islam.html.

Moore, T.E. (1996). Scientific consensus and expert testimony: Lessons from the Judas Priest trial. Retrieved July 5, 2012, from http://www.csicop.org/si/show/scientific_consensus_and_expert_testimony.

Morris, H.M. (n.d.). Evolution, thermodynamics and entropy. Retrieved August 27th, 2012, from http://www.icr.org/article/51/247/.

Mostert, M.P. (2001). Facilitated communication since 1995: A review of published studies. Journal of Autism and Developmental Disorders, 31, 287-313.

Movromatis, A. (1986). Hypnagogia: The unique state of consciousness between wakefulness and sleep. London: Routledge & Kegan Paul Ltd. ISBN: 978-0710202826.

Murra, D., & Di Lazzaro, P. (2010). Sight and brain: An introduction to the visually misleading image. Poster presented as the Proceedings of the International Workshop on the Scientific approach to the Acheiropoietos Images, ENEA Frascati, Italy, 4-6 May.

Muzur, A., Pace-Schott, E.F., & Hobson, J.A. (2002). The prefrontal cortex in sleep. TRENDS in Cognitive Sciences, 6, 475-481.

Myers, D.G., & Lamm, H. (1976). The group polarisation phenomenon. Psychological Bulletin, 83, 602-627. doi: 10.1037/0033-2909.83.4.602.

Myers, P.Z. (2008). Entropy and evolution. Retrieved August 27th, 2012, from http://scienceblogs.com/pharyngula/2008/11/10/entropy-and-evolution/.

Nelson, K.B., & Bauman, M.L. (2003). Thimerosal and autism? Pediatrics, 111, 674-679. doi: 10.1542/peds.111.3.674.

Newbrook, M. (2005). Reverse speech. Retrieved July 5, 2012, from http://www.csicop.org/sb/show/reverse_speech/.

Newbrook, M., & Curtain, J.M. (1998). Oates' theory of reverse speech: A critical examination. International Journal of Speech Language and the Law, 5, 174-192.

Ng, F. (2000). Tinky Winky speaks! Mother sues! Retrieved July 5, 2012, from http://www.hollywood.com/news/EXTRA_Tinky_Winky_Speaks_Mother_Sues/312207.

NICE: Guideline development methods - Reviewing and grading the evidence. (2005). Retrieved August 6th, 2012, from http://www.nice.org.uk/niceMedia/pdf/GDM_Chapter7_0305.pdf.

Nickerson, R.S. (1998). Confirmation bias: A ubiquitous phenomenon in many guises. Review of General Psychology, 2, 175-220.

Nicolas, S., Collins, T., Gounden, Y., & Roediger, H.L. (2011). The influence of suggestibility on memory. Consciousness and Cognition, 20, 399-400. doi: 10.1016/j.concog.2010.10.019.

Nielsen, R. (2006). Limits of opinion polls: Comparing MAAR & Roper Surveys. Retrieved July 9, 2012, from http://www.ufodigest.com/news/1106/opinionpolls.html.

Nishiyama, Y. (2010). Pattern matching probabilities and paradoxes as a new variation on Penney's coin game. International Journal of Pure and Applied mathematics, 59, 357-366.

Nordland, R. (2009). Iraq Swears by bomb detector U.S. sees as useless. Retrieved June 17, 2012, from http://www.nytimes.com/2009/11/04/world/middleeast/04sensors.html.

Nordquist, R. (n.d.). Confirmation bias. Retrieved July 1, 2012, from http://grammar.about.com/od/c/g/Confirmation-Bias.htm.

Novella, R. (2000). The power of coincidence: some notes on "psychic" predictions. Retrieved June 27, 2012, from

Bibliography and further reading

http://www.quackwatch.com/04ConsumerEducation/coincidence.html.

Novella, R. (2001). Pseudoscience in the rubble. Retrieved July 5, 2012, from http://www.theness.com/index.php/911-pseudoscience-in-the-rubble/.

Novella, S. (2007). Perpetual idiocy. Retrieved August 8th, 2012, from http://theness.com/neurologicablog/index.php/perpetual-idiocy/.

Novella, S. (2008). Facilitated testimony in the courtroom. Retrieved July 1, 2012, from http://theness.com/neurologicablog/index.php/facilitated-testimony-in-the-courtroom/.

Novella, S. (2008). More on FC in the courtroom and the Oakland case. Retrieved July 1, 2012, from http://theness.com/neurologicablog/index.php/more-on-fc-in-the-courtroom-and-the-oakland-case/.

Novella, S. (2009). Acupuncture does not work for back pain. Retrieved August 8th, 2012, from http://www.sciencebasedmedicine.org/index.php/acupuncture-does-not-work-for-back-pain/.

O'Brien, B., & Ellsworth, P.C. (2006). Confirmation bias in criminal interviews. 1st Annual Conference on Empirical Legal Studies Paper.

Oates, D.J. (2001). What is reverse speech? Retrieved July 5, 2012, from http://www.reversespeech.com/words_of_creation.htm.

Oates, D.J. (n.d.). Bill Clinton reversals. Retrieved July 6, 2012, from http://www.reversespeech.com/clinton.htm.

Oates, D.J. (n.d.). O.J. Simpson reversals. Retrieved July 6, 2012, from http://www.reversespeech.com/ojsimpson.htm.

One million dollar paranormal challenge. (2008). Retrieved June 23, 2012, from http://www.randi.org/site/index.php/1m-challenge.html.

Ongley, P. A. (1948). New Zealand diviners. The New Zealand journal of Science and Technology. 30, 38-54.

Oxford Centre for Evidence-based Medicine - Levels of evidence. (2009). Retrieved August 6th, 2012, from http://www.cebm.net/index.aspx?o=1025.

Palfreman, J. (2012). The dark legacy of FC. Evidence-Based Communication Assessment and Intervention. iFirst Article, 1-4. doi: 10.1080/17489539.2012.688343.

Parents cleared in sex case file suit. (2008). Retrieved July 1, 2012, from http://www.freep.com/apps/pbcs.dll/article?AID=/20080912/NEWS03/809120414.

Parker, S.K., Schwartz, B., Todd, J., & Pickering, L.K. (2004). Thimerosal-containing vaccines and autistic spectrum disorder: A critical review of published original data. Pediatrics, 114, 793-804. doi: 10.1542/peds.2004-0434.

Paulos, J.A. (2001). Innumeracy: Mathematical illiteracy and its consequences. New York: Hill and Wang. ISBN: 978-0679726012.

Paulovic, M. (2010). Backward masking. Unpublished essay, Heriot-Watt University.

Payne, K.J. (2009). We see what isn't there. Retrieved July 3, 2012, from http://kjpayne.com/2009/08/we-see-what-isnt-there/.

Peterson, T., Kaasa, S.O., & Loftus, E.F. (2009). Me too!: Socual modelling influences on early autobiographical memories. Applied Cognitive Psychology, 23, 267-277. doi: 10.1002/acp.1455.

Phillips, M.D. (n.d.). Satan in the Smoke? A photojournalist's 9/11 story. Retrieved July 5, 2012, from http://www.markdphillips.com/.

Pigliucci, M. (2010). Nonsense on Stilts – How to Tell Science from Bunk. Chicago: The University of Chicago Press. ISBN: 978-0226667867.

Plait, P. (2002). Bad Astronomy. New York: John Wiley and Sons. ISBN: 0-471409766.

Plait, P. (2004). Are there artifacts on Mars? Bizarre image analysis. Retrieved July 3, 2012, from http://www.badastronomy.com/bad/misc/hoagland/artifacts.html.

Plait, P. (2004). Face to face. Retrieved July 3, 2012, from http://www.badastronomy.com/bad/misc/hoagland/face.html.

Plait, P. (2008). Death from the Skies. New York: Viking Penguin. ISBN: 978-0670019977.

Pope. K.S. (1996). Memory, abuse, and science: Questioning claims about the False Memory Syndrome epidemic. American Psychologist, 51, 957-974. doi: 10.1037/0003-066X.51.9.957.

Bibliography and further reading

Population Estimates for UK, England and Wales, Scotland and Northern Ireland, mid-2010. (2011). Retrieved June 27, 2012, from http://www.ons.gov.uk/ons/rel/pop-estimate/population-estimates-for-uk--england-and-wales--scotland-and-northern-ireland/mid-2010-population-estimates/index.html.

Posner, G.P. (2000). The face behind the "face" on Mars: A skeptical look at Richard C. Hoagland. Retrieved July 3, 2012, from http://www.gpposner.com/Hoagland.html.

Pratkanis, A.R. (1992). The cargo-cult science of subliminal persuasion. Retrieved July 5, 2012, from http://www.csicop.org/si/show/cargo-cult_science_of_subliminal_persuasion.

Pronin, E., & Kugler, M.B. (2007). Valuing thoughts, ignoring behaviour: The introspection illusion as a source of the bias blind spot. Journal of Experimental Social Psychology, 43, 565-578. doi: 10.1016/j.jesp.2006.05.011.

Pronin, E., Gilovich, T., & Ross, L. (2004). Objectivity in the eye of the beholder: Divergent perceptions of bias in the self verses others. Psychological Review, 111, 781-799. doi: 10.1037/0033-295X.111.3.781.

Pronin, E., Lin, D.Y., & Ross, L. (2002). The bias blind spot: perceptions of bias in self versus others. Personality and Social Psychology Bulletin, 28, 369-381. doi: 10.1177/0146167202286008.

Quinton, J. (2012). The problem with Ad Hoc hypotheses (Bayes' theorem and coin flips). Retrieved August 10th, 2012, from http://deusdiapente.blogspot.co.uk/2012/03/problem-with-ad-hoc-hypotheses-bayes.html.

Radford, B. (2010). Scientific Paranormal Investigations – How to Solve Unexplained Mysteries. Corrales, N.M.: Rhombus Publishing Company. ISBN: 978-0936455112.

Radford, B. (2012). Airplane, UFO or Venus. Retrieved July 9, 2012, from http://news.discovery.com/human/airplane-ufo-or-venus-120420.html.

Radford, B., & Nickell, J. (2006). Lake Monster Mysteries – Investigating the World's Most Elusive Creatures. Lexington, Kentucky: The University Press of Kentucky. ISBN: 978-0813123943.

Radio receiver cross modulation. (n.d.). Retrieved July 6, 2012, from http://www.radio-electronics.com/info/rf-technology-design/receiver-overload/crossmodulation.php.

Radish, R.O.T. (2007). Songs accused of Satanic backmasking. Retrieved July 5, 2012, from http://music.yahoo.com/blogs/yradish/songs-accused-of-satanic-backmasking.html.

Ramsland, K. (n.d.). Shadow of a doubt: The Clarence Elkins story. Retrieved June 23, 2012, from http://www.trutv.com/library/crime/criminal_mind/forensics/ff313_clarence_elkins/4.html.

Randi, J. (1979). A controlled test of dowsing abilities. The Skeptical Inquirer the Zetetic, 4, 16-20.

Randi, J. (1982). Flim-Flam! Psychics, ESP, Unicorns, and Other Delusions (Reprint edition). Amherst, New York: Prometheus Books. ISBN: 978-0879751982.

Randi, J. (1995). An encyclopedia of claims, frauds, and hoaxes of the occult and supernatural. Geller, Uri. Retrieved June 28. 2012, from http://www.randi.org/encyclopedia/Geller,%20Uri.html.

Randi, J. (2005). Fakers and innocents: The one million dollar challenge and those who try for It. Retrieved June 23, 2012, from http://www.csicop.org/si/show/fakers_and_innocents_the_one_million_dollar_challenge_and_those_who_try_for/.

Raudive, K. (1971). Breakthrough: An amazing experiment in electronic communication with the dead. Retrieved July 6, 2012, from http://antiworld.se/project/references/texts/Breakthrough.Extract.pdf.

Redwood, D. (1995). Quantum Healing: Interview with Deepak Chopra. Retrieved August 27th, 2012, from http://www.healthy.net/scr/interview.aspx?Id=167.

Religious icons. The miracles just keep coming. (n.d.). Retrieved July 3, 2012, from http://www.yoism.org/?q=node/129.

Revoir, P. (2009). OK Derren, now tell us how you really did it: Experts pour scorn on illusionist's explanation of lottery stunt. Retrieved June 25, 2012, from http://www.dailymail.co.uk/tvshowbiz/article-1212839/Okay-Derren-Brown-tell-REALLY-did-Experts-pour-scorn-illusionists-explanation-lottery-stunt.html.

Bibliography and further reading

Richards, A., French, C.C., Harris, P. (1998). Mistakes around the clock: Errors in memory for orientation of numerals. The Journal of Psychology: Interdisciplinary and Applied, 132, 42-46. doi: 10.1080/00223989809599263.

Richmond, S.J., Brown, S.R., Campion, P.D., Porter, A.J., Moffett, J.A., Jackson, D.A., Featherstone, V.A., & Taylor, A.J. (2009). Therapeutic effects of magnetic and copper bracelets in osteoarthritis: A randomised placebo-controlled crossover trial. Complementary Therapies in Medicine, 17, 246-256.

Ridpath, I. (2010). Astronomical causes of UFOs. Retrieved August 8th, 2012, from http://www.ianridpath.com/ufo/astroufo1.htm.

Rips, L.J. (2002). Circular reasoning. Cognitive Science, 26, 767-795.

Robbins, S. (2010). Why you can't believe anyone who says they know the size and distance of a UFO. Retrieved July 9, 2012, from http://pseudoastro.wordpress.com/2010/08/13/why-you-cant-believe-anyone-who-says-they-know-the-size-and-distance-of-a-ufo/.

Robbins, S. (2012). That was fast: Neutrinos can't travel faster than light (so far). Retrieved July 20, 2012, from http://pseudoastro.wordpress.com/tag/pons-and-fleischmann/.

Robinson, B.A. (2001). Christian urban legends: Backmasking on records: real, of hoax? Retrieved July 5, 2012, from http://www.religioustolerance.org/chr_cul5.htm/.

Robinson, L. (n.d.). UFOs You've been hoodwinked! Mistakes and misidentifications in UFO research. Retrieved July 9, 2012, from http://midimagic.sgc-hosting.com/ufopage.htm.

Ronson, J. (2004). The Men who Stare at Goats. London: Picador. ISBN: 0-330375474.

Rosenthal, R. (2002). Covert communication in classrooms, clinics, courtrooms, and cubicles. American Psychologist, 57, 839-849.

Rosenthal, R. (2003). Covert communication in laboratories, classrooms, and the truly real world. Current Directions in Psychological Science, 12, 151-154. doi: 10.1111/1467-8721.t01-1-01250.

Ross, L., Greene, D., & House, P. (1977). The "false consensus effect": An egocentric bias in social perception and attribution processes.

Journal of Experimental Social Psychology, 13, 279-301. doi: 10.1016/0022-1031(77)90049-X.

Rowland, I. (2008). The Full Facts book of Cold Reading (Forth Edition). London: Full Facts Books. ISBN: 978-0955847608.

Roy, A. (2005). The glorious name "ALLAH" appeared on the waves of tsunami! : A response. Retrieved July 5, 2012, from http://mukto-mona.com/rationalism/tsunami_allah.htm.

Sabbagh, K. (2009). The Hair of the Dog. London: John Murray. ISBN: 978-1848540880.

Sagan, C. (2006). The Demon-Haunted World. New York: Ballantine Books. ISBN: 978-0345409461.

Schacter, D.L. (1996). Searching for memory: The brain, the mind, and the past. New York: Basic Books. ISBN: 978-0465075522.

Schulz, H. (2008). Rethinking sleep analysis: Comment on the AASM manual for the scoring of sleep and associated events. Journal of Clinical Sleep Medicine, 4, 99-103.

Seargent, D.A.J. (2010). Weird astronomy: Tales of unusual, bizarre, and other hard to explain observations. London: Springer. ISBN: 978-1441964236.

Seitz, A.R., Nanez, J.E., Holloway, S.R., Koyama, S., & Watanabe, T. (2005). Seeing what is not there shows the costs of perceptual learning. Proceedings of the National Academy of Sciences, 102, 9080–9085.

Seitz, A.R., Yamagishi, N., Werner, B., Goda, N., Kawato, M., & Watanabe, T. (2005). Task-specific disruption of perceptual learning. Proceedings of the National Academy of Sciences, 102, 14895–14900.

Sengupta, K. (2010). Head of bomb detector company arrested in fraud investigation. Retrieved June 17, 2012, from http://www.independent.co.uk/news/uk/crime/head-of-bomb-detector-company-arrested-in-fraud-investigation-1876388.html.

September 11 news.com. Mysteries —Faces & crosses. (n.d.). Retrieved July 5, 2012, from http://www.september11news.com/Mysteries1.htm.

Shaw. G.M. (1991). The admissibility of hypnotically enhanced testimony in criminal trials. Marquette Law Review, 75, 1-77.

Bibliography and further reading

Shermer, M. (2005). Turn me on, dead man. Retrieved July 5, 2012, from http://www.scientificamerican.com/article.cfm?id=turn-me-on-dead-man.

Shermer, M. (2007). Why people believe weird things: Pseudoscience, Superstition and other confusions of our time. London: Souvenir Press. ISBN: 978-0285638037.

Shostak, S. (2009). Confessions of an Alien Hunter. Washington, D.C.: National Geographic. ISBN: 978-1426203923.

Simons, D.J., & Chabris, C.F. (2011). What people believe about how memory works: A representative survey of the U.S. population. PLoS ONE, 6(8): e22757. doi:10.1371/journal.pone.0022757.

Sina, A. (n.d.). The miracles of Allah. Retrieved July 5, 2012, from http://www.faithfreedom.org/Articles/sina/miracles_of_allah.htm.

Singh, S. (2004). Big Bang. London: Harper Perennial. ISBN: 978-0007152520.

Singh, S., & Edzard, E. (2008). Trick or Treatment. London: Bantam Books. ISBN: 978-0552157629.

Skeptico. (2010). Mike Adams fails again. Retrieved June 28, 2012, from http://skeptico.blogs.com/skeptico/2010/02/mumps-new-jersey-77-vaccinated-get-infected-mike-adams-fails.html.

Sleep paralysis and associated hypnagogic and hypnopompic experiences. (n.d.). Retrieved July 7, 2012, from http://watarts.uwaterloo.ca/~acheyne/S_P.html.

Sleep paralysis. (2006). Retrieved July 7, 2012, from http://www.h2g2.com/approved_entry/A6092471.

Sleep paralysis: Medical explanation. (2010). Retrieved July 7, 2012 from http://medchrome.com/patient/sleep-problems/sleep-paralysis-alien-abduction-demon-or-physiological/.

Spanos, N. P., Cross, P. A., Dickson, K., & DuBreuil, S. C. (1993). Close encounters: An examination of UFO experiences. Journal of Abnormal Psychology, 102, 624-632.

Staehle, H.J., Koch, M.J., & Pioch, T. (2005). Double-blind study on materials testing with applied kinesiology. Journal of Dental Research, 84, 1066-1069.

Stibich, M. (2009). The stages of sleep. Retrieved July 7, 2012, from http://longevity.about.com/od/sleep/a/sleep_stages.htm.

Styer, D.F. (2008). Entropy and evolution. American Journal of Physics, 76, 1031-1033. doi: 10.1119/1.2973046.

Sutton, A.J., Duval, S.J., Tweedie, R.L., Abrams, K.R., & Jones, D.R. (2000). Empirical assessment of effect of publication bias on meta-analyses. British Medical Journal, 320, 1574-1577. doi: 10.1136/bmj.320.7249.1574.

Svoboda, E. (2007). Faces, faces everywhere. Retrieved July 3, 2012, from http://www.nytimes.com/2007/02/13/health/psychology/13face.html?_r=1.

Taber, C.S., & Lodge, M. (2006). Motivated skepticism in the evaluation of political beliefs. American Journal of Political Science, 50, 755-769. doi: 10.1111/j.1540-5907.2006.00214.x.

Takahashi, K. (2007). The effect of subliminal messages and suggestions on memory: Isolating the placebo effect. Electronic Theses, Treatises and Dissertations. Paper 1732.

Tavris, C., & Aronson, E. (2008). Mistakes Were Made (but not by me). London: Pinter & Martin Ltd. ISBN: 978-1905177219.

Tenety, E. (2011). May 21, 2011: Harold Camping says the end is near. Retrieved June 23, 2012, from http://onfaith.washingtonpost.com/onfaith/undergod/2011/01/may_21_2011_harold_camping_says_the_end_is_near.html.

The book test mind reading trick explained by scam school. (n.d.). Retrieved June 28, 2012, from http://www.learnmagictricks.org/video.php?v=25880.

The chances of winning the UK national lottery. (n.d.). Retrieved June 25, 2012, from http://playlotto.org.uk/lottery/uklottery_odds.html.

The jury is in: No free energy from Steorn. (2009). Retrieved August 31st, 2012, from http://dispatchesfromthefuture.com/2009/06/the_jury_is_in_no_free_energy_from_steor.html.

The talking Fisher Price doll that 'preaches pro-Islamic message to toddlers'. (2008). Retrieved July 5, 2012, from http://www.dailymail.co.uk/news/article-1077316/The-talking-Fisher-Price-doll-preaches-pro-Islamic-message-toddlers.html.

Bibliography and further reading

Thompson, K. (1998). Moral Panics. London: Routledge. ISBN: 978-0415119764.

Thorne, S.B., & Himelstein, P. (1984). The role of suggestion in the perception of satanic messages in rock-and-roll recordings. The Journal of Psychology, 116, 245-248.

Toothpick acupuncture better than drugs. (2009). Retrieved August 8th, 2012, from http://www.mirror.co.uk/news/uk-news/toothpick-acupuncture-better-than-drugs-393502.

TransCommunication white paper with emphasis on electronic voice phenomena. (2011). Retrieved July 6, 2012, from http://atransc.org/ATransC%20White%20Paper%20on%20EVP.pdf.

UK national lotto sales figures (n.d.). Retrieved June 25, 2012, from http://lottery.merseyworld.com/Sales_index.html.

UK UFO sightings: Bromsgrove,Worcestershire-3rd June 2010. (2010). Retrieved July 9, 2012, from http://www.uk-ufo.co.uk/bromsgroveworcestershire-3rd-june-2010/.

United States census bureau population clock projection (2012). Retrieved June 27, 2012, from http://www.census.gov/population/www/popclockus.html.

Unmasking the face on Mars. (2001). Retrieved July 3, 2012, from http://science.nasa.gov/science-news/science-at-nasa/2001/ast24may_1/.

Urban legends running rampant / From smoky Satan to Nostradamus. (2001). Retrieved July 5, 2012, from http://www.sfgate.com/news/article/Urban-legends-running-rampant-From-smoky-Satan-2877101.php.

Vetrugno, R., & Montagna, P. (2011). Sleep-to-wake transition movement disorders. Sleep Medicine, 12, 511-516. doi: 10.1016/j.sleep.2011.10.005.

Virgin Mary grilled cheese sells for $28,000. (2004). Retrieved July 5, 2012, from http://www.msnbc.msn.com/id/6511148/.

Vokey, J.R., & Read, D. (1985). Subliminal messages: Between the Devil and the media. American Psychologist, 40, 1231-1239.

Voss, J.L., Federmeier, K.D., & Paller, K.A. (2011). The potato chip really does look like Elivs! Neural hallmarks of conceptual processing associated with finding novel shapes subjectively meaningful. Cerebral Cortex, bhr315. doi: 10.1093/cercor/bhr315.

Wagner, S. (n.d.). How to practice automatic writing. Retrieved July 1, 2012, from http://paranormal.about.com/cs/automaticwriting/ht/auto_writing.htm.

Wakefield, A.J., Murch, S.H., Anthony, A., Linnell, J., Casson, D.M., Malik, M., et al. (1998). Ileal-lymphoid-nodular hyperplasia, non-specific colitis, and pervasive developmental disorder in children. The Lancet, 351, 637–641. (Note: This paper has subsequently been withdrawn due to evidence that fraudulent data was used.)

Wallace, A. (2009). An epidemic of fear: How panicked parents skipping shots endanger us all. Retrieved June 17, 2012, from http://www.wired.com/magazine/2009/10/ff_waronscience.

Wason, P.C. (1960). On the failure to eliminate hypotheses on conceptual task. Quarterly Journal of Experimental Psychology, 12, 129-140. doi: 10.1080/17470216008416717.

Wason, P.C. (1968). Reasoning about a rule. Quarterly Journal of Experimental Psychology, 20, 273-281. doi: 10.1080/14640746808400161.

Wason, P.C., & Shapiro, D. (1971). Natural and contrive experience in a reasoning problem. Quarterly Journal of Experimental Psychology, 23, 63-71.

Watson, S. (2002). How EVP works. Retrieved July 6, 2012, from http://science.howstuffworks.com/science-vs-myth/afterlife/evp.htm.

Weaver, K., Garcia, S.M., Schwarz, N., & Miller, D.T. (2007). Inferring the popularity of an opinion from its familiarity: A repetitive voice can sound like a chorus. Journal of Personality and Social Psychology, 92, 821-833. doi: 10.1037/0022-3514.92.5.821.

Wegner, D.M., Fuller, V.A., & Sparrow, B. (2003). Clever hands: Uncontrolled intelligence in facilitated communication. Journal of Personality and Social Psychology, 85, 5-19. doi: 10.1037/0022-3514.85.1.5.

Weiner, I.B., & Hess, A.K. (2006). The handbook of forensic psychology, third edition. Hoboken, New Jersey: John Wiley & Sons, Inc.

Bibliography and further reading

What is applied kinesiology? (2005). Retrieved June 29, 2012, from http://www.icak.com/.

Wheeler, D.L., Jacobson, J.W., Paglieri, R.A., & Schwartz, A.A. (1993). An experimental assessment of facilitated communication. Mental Retardation, 31, 49-59.

Whitehouse, D. (2001). NASA: No face – honest. Retrieved July 3, 2012, from http://news.bbc.co.uk/1/hi/sci/tech/1351319.stm.

Wilkins, J.S. (2004). Spontaneous generation and the origin of life. Retrieved July 13, 2012, from http://www.talkorigins.org/faqs/abioprob/spontaneous-generation.html.

Wilson, J. (2007). Scientific laws, hypothesis, and theories. Retrieved July 13, 2012, from http://wilstar.com/theories.htm.

Wiseman, R. (2007). Quirkology. London: MacMillan. ISBN: 978-0230702158.

Wiseman, R. (2009). :59 Seconds. London: MacMillan. ISBN: 978-0230744295.

Wiseman, R., & Greening, E. (2005). 'It's still bending': Verbal suggestion and alleged psychokinetic ability. British Journal of Psychology, 96, 115-127.

Wiseman, R., & Schlitz, M. (1997). Experimenter effects and the remote detection of staring. Retrieved July 1, 2012, from http://findarticles.com/p/articles/mi_m2320/is_n3_v61/ai_20749205/.

Wolford, G., Newman, S.E., Miller, M.B., & Wig, G.S. (2004). Searching for patterns in random sequences. Canadian Journal of Experimental Psychology, 58, 221-228.

Wong, D. (2010). Explaining ghost sightings (part 1 – pareidolia). Retrieved July 3, 2012, from http://warforscience.wordpress.com/2010/03/16/explaining-ghost-sightings-part-1-%E2%80%93-pareidolia/.

Wood, W., Lundgren, S., Ouellette, J.A., Busceme, S., & Black, T. (1994). Minority influence: A meta-analytic review of social influence processes. Psychological Bulletin, 115, 323-345.

Wright, C. (2008). Toy doll sparks complaints. Retrieved July 5, 2012, from http://www.newson6.com/global/story.asp?s=9149557.

Wurlich, B. (2005). Unproven techniques in allergy diagnosis. Journal of investigational allergology and clinical immunology, 15, 86–90.

Young, L.K. (n.d.). Primacy effect on impression formation. Retrieved July 1, 2012, from http://psych.fullerton.edu/mbirnbaum/psych466/ykl/report.htm.

Yudkowsky, E.S. (2003). An intuitive explanation of Bayesian reasoning. Retrieved August 10th, 2012, from http://yudkowsky.net/rational/bayes/.

Zhu, B., Chen, C., Loftus, E.F., He, Q., Chen, C., Lei, X., Lin, C., & Dong, Q. (2012). Brief exposure to misinformation can lead to long-term false memories. Applied Cognitive Psychology, 26, 301-307. doi: 10.1002/acp.1825.

Zhu, B., Chen, C., Loftus, E.F., Lin, C., & Dong, Q. (2010). Treat and trick: a new way to increase false memory. Applied Cognitive Psychology, 24, 1199-1208. doi: 10.1002/acp.1637.

Zhu, H., Huberman, B., & Luon, Y. (2012). To switch or not to switch: understanding social influence in online choices. Proceedings of the 2012 ACM annual conference on Human Factors in Computing Systems, 2257-2266. doi: 10.1145/2207676.2208383.

Zimmerman Jones, A. (n.d.). Quantum physics overview. Retrieved August 27th, 2012, from http://physics.about.com/od/quantumphysics/p/quantumphysics.htm.

Printed in Great Britain
by Amazon